04-HHE-5

# OPTIMIZING FILTRATION AND DISINFECTION SYSTEMS WITH A RISK-BASED APPROACH

by:
Frank J. Loge
Jeannie L. Darby
Timothy R. Ginn
Mary Kay Camarillo
University of California, Davis

*2010*

WERF

The Water Environment Research Foundation, a not-for-profit organization, funds and manages water quality research for its subscribers through a diverse public-private partnership between municipal utilities, corporations, academia, industry, and the federal government. WERF subscribers include municipal and regional water and wastewater utilities, industrial corporations, environmental engineering firms, and others that share a commitment to cost-effective water quality solutions. WERF is dedicated to advancing science and technology addressing water quality issues as they impact water resources, the atmosphere, the lands, and quality of life.

For more information, contact:
Water Environment Research Foundation
635 Slaters Lane, Suite 300
Alexandria, VA 22314-1177
Tel: (703) 684-2470
Fax: (703) 299-0742
www.werf.org
werf@werf.org

This report was co-published by the following organization.

IWA Publishing
Alliance House, 12 Caxton Street
London SW1H 0QS, United Kingdom
Tel: +44 (0) 20 7654 5500
Fax: +44 (0) 20 7654 5555
www.iwapublishing.com
publications@iwap.co.uk


Library of Congress Catalog Card Number: 2008943801
IWAP ISBN: 978-1-84339-280-4/1-84339-280-1



University of California, Davis

The research on which this report is based was developed, in part, by the United States Environmental Protection Agency (EPA) through Cooperative Agreement No. CR83155901-1 with the Water Environment Research Foundation (WERF). However, the views expressed in this document are not necessarily those of the EPA and EPA does not endorse any products or commercial services mentioned in this publication. This report is a publication of WERF, not EPA. Funds awarded under the Cooperative Agreement cited above were not used for editorial services, reproduction, printing, or distribution.

# ACKNOWLEDGMENTS


## Research Team

**Principal Investigator:**
Frank J. Loge, Ph.D., *University of California, Davis*

**Co-Principal Investigators:**
Jeannie L. Darby, Ph.D., *University of California, Davis*

Timothy R. Ginn, Ph.D., *University of California, Davis*

**Project Team Member:**
Mary Kay Camarillo, *University of California, Davis*

## Project Subcommittee

William L. Cairns, Ph.D.
*Trojan Technologies*

John Chorlog
*Miami-Dade Water & Sewer Department*

Charles I. Noss, Sc.D.
*U.S. Environmental Protection Agency*

Pierre Payment, Ph.D.
*Institut national de la recherche scientifique -Institut Armand-Frappier*

Mary K. Stinson
*U.S. Environmental Protection Agency*

Elliott Whitby, Ph.D.
*Calgon Carbon Corporation*

Moyasar T. Yahya, MS
*Charles Abbott Associates*

## Water Environment Research Foundation Staff

**Director of Research:** Daniel M. Woltering, Ph.D.
**Senior Program Manager:** Lola Olabode, M.P.H

# ABSTRACT AND BENEFITS

**Abstract:**

Treated wastewater effluents contain high concentrations of particles; many of these particles are large (with diameters greater than 100 μm) and contain bacterial cells enmeshed in organic matter. Microorganisms occluded in wastewater particles are difficult to inactivate in chlorine disinfection systems, as chlorine ions must first diffuse through the macro- and microscopic pore spaces prior to inactivating occluded microorganisms. Particle occlusion results in tailing, where reduced inactivation rates occur even with increasing disinfectant doses. Reduction of occluded microorganisms in plant effluents is accomplished using filtration to remove the total number of particles, and disinfection to ensure that intra-particle chlorine concentrations are sufficient to inactivate occluded microorganisms. In addition to inactivating dispersed microorganisms, treatment systems should be designed and operated to reduce concentrations of microorganisms occluded in wastewater particles that may pose a health risk in post treated waters.

In this project, a systematic approach was developed to co-optimize filtration and chemical disinfection systems to collectively reduce the concentration of occluded microorganisms in treated effluents to acceptable levels. The optimization process was successfully applied to wastewater samples collected from seven facilities, each with different treatment trains. Particle guidelines were developed that ensure concentrations of occluded microorganisms were sufficiently reduced. In addition, a range of operating conditions (particle counts and chlorine doses) was identified that resulted in acceptable treatment. Particle guidelines and the optimization approach were based on the existing regulatory framework for indicator organisms. Extension of the current approach to a pathogen basis was considered, but current data are insufficient to adopt such a procedure. However, preliminary results suggest that intra-particle chlorine concentrations that are sufficient to inactivate indicator organisms may not be adequate to sufficiently reduce concentrations of occluded pathogens.

**Benefits:**

This research project was completed to assist professionals working in the treatment industry to address the disinfection of microorganisms occluded in wastewater particles, and the role of filtration in removing such microorganisms. Specifically, it:

- Allows managers of treatment facilities to evaluate their own systems using particle guidelines and the step-by-step optimization process, allowing them to consider alternative configurations of their filtration and disinfection systems.
- Allows managers of treatment facilities to evaluate results obtained for facilities with similar treatment trains and process control parameters.
- Uses particle size distribution (PSD) data in a meaningful manner that is connected to the performance of the disinfection system, and allows use of PSD as a process control parameter.
- Develops particle guidelines for wastewater effluents, which are based on the existing regulatory framework for indicator organisms.

- Describes a methodology for extending the concept of particle guidelines and optimization protocols to a pathogen basis, should the necessary information for pathogens become available.

**Keywords:** Wastewater, chlorine disinfection, filtration, particle size distribution, diffusion.

# TABLE OF CONTENTS

# LIST OF TABLES

# LIST OF FIGURES

# LIST OF ACRONYMS AND NOMENCLATURE

**Acronyms**

| | |
|---|---|
| CA DPH | California Department of Public Health |
| CFU | Colony forming unit |
| COD | Chemical oxidizing demand |
| CSDLAC | County Sanitation Districts of Los Angeles County |
| CT | Chlorine residual concentration multiplied by contact time (mg*min/L) |
| CWRC | California Water Recycling Criteria |
| IU | Infectious units |
| MCRT | Mean cell residence time |
| MPN | Most probable number |
| PFU | Plaque forming unit |
| PSD | Particle size distribution |
| rpm | Revolutions per minute |
| U.S. EPA | United States Environmental Protection Agency |
| UV | Ultraviolet |

**Nomenclature**

| | |
|---|---|
| $Bi$ | Mass transfer associated Biot number |
| $c$ | Intra-particle aqueous concentration (mg/L) |
| $c_0$ | Initial uniformly distributed intra-particle aqueous concentration (mg/L) |
| $c_{bulk}$ | Bulk solution aqueous concentration (mg/L) |
| $c_j$ | Intra-particle aqueous concentration in the jth particle size class (mg/L) |
| $c_{ref}$ | Maximum concentration observed initially in the reacted or dissolved phase (mg/L$^3$) |
| $c_{SM}$ | Soluble organic in the bulk fluid concentration (mg/L) |
| $C$ | Dimensionless intra particle dissolved concentration |
| $C_{bulk}$ | Dimensionless bulk concentration |
| $C_{bulk,in}$ | Dimensionless initial bulk concentration |
| $C_{in}$ | Dimensionless initial intra-particle aqueous concentration |
| $C_{SM}$ | Dimensionless soluble organic matter concentration in the bulk fluid |
| $cv$ | Coefficient of variation |
| $d$ | Number of microorganisms or viruses that an individual is exposed to (microbial units/(person*exposure)) |
| $D$ | Bulk fluid diffusion coefficient (μm$^2$/min) |
| $D_a$ | Intra-particle diffusion coefficient (μm$^2$/min) |
| $F$ | Fraction of particles that have occluded microorganisms (number of particles with occluded microorganisms/total number of particles) |
| $F_j$ | F (see above) in the jth particle size class |
| $k$ | Intra-particle aqueous phases decay rate (min$^{-1}$) |
| $k'$ | Bulk aqueous phases decay rate (min$^{-1}$) |
| $k_f$ | Surface oxidation reaction coefficient (min$^{-1}$) |
| $k_m$ | Mass-transfer coefficient across the particle boundary (μm /min) |
| $k_i$ | Inactivation rate coefficient (L/mg/min) |

| | |
|---|---|
| $k_r$ | Desorption rate coefficient ($min^{-1}$) |
| $k_{SM}$ | Decay rate for the soluble matter ($min^{-1}$) |
| $K_d$ | Ratio of the sorption rate to the desorption rate |
| $K_p$ | Equilibrium partition coefficient at the particle surface |
| $N(t)$ | Concentration of microorganisms following exposure to a disinfectant for time t (microbial units/L) |
| $N_0$ | Initial concentration of particles with diffusion limitations that contain occluded microorganisms (microbial units/100 mL) |
| $N_{50}$ | Median infectious dose (microbial units) |
| $N_d$ | Initial concentration of dispersed microorganisms (microbial units /L) |
| $N_p$ | Particle guideline, which is the maximum allowable concentration of particles in the effluent (total number of particles/L) |
| $nts$ | Number of time steps |
| $P$ | Number of pathogens occluded per particle in the sub-domain of particles that contain one or more occluded pathogens (microbial units/number of particles with occluded microorganisms) |
| $P_a$ | Annual probability of infection or illness due to pathogens |
| $P_{bp}$ | Probability of infection or illness in the beta-Poisson model |
| $P_e$ | Probability of infection or illness in the exponential model |
| $P_s$ | Probability of infection or illness due to the presence of pathogens in water, single exposure event |
| $r$ | Spatial index along the radius (μm) |
| $r_e$ | Variable in the exponential risk assessment model |
| $R$ | Reference particle radius (μm) |
| $R_i$ | Particle radius for the ith size class (μm) |
| $s$ | Intra-particle sorbed mass per mass of particle (mg/mg) |
| $s_0$ | Uniformly distributed initial intra-particle sorbed mass per mass of particle (mg/mg) |
| $S$ | Dimensionless intra-particle sorbed concentration |
| $S_{in}$ | Dimensionless initial intra-particle sorbed concentration |
| $t$ | Time (min) |
| $V_c$ | Volume of water consumed per person per day (L/person*exposure) |
| $V$ | Volume of the reactor used in the disinfection tests (L) |
| | |
| $\varepsilon$ | Ratio of intra-particle aqueous volume of a particle to the extra-particle aqueous volume |
| $\lambda$ | Characteristic parameter of the Poisson probability distribution |
| $\eta$ | Relative importance of intra-particle desorption to mass transfer at the particle surface |
| $\theta$ | Particle porosity, ratio of intra-particle aqueous volume to the total particle volume |
| $\xi$ | Dimensionless length |
| $\rho$ | Mass of sorbent per total particle volume (mg/L) |
| $\rho_p$ | Particle wet density (mg/L) |
| $\rho_s$ | Density of solid particle material (mg/L) |
| $\rho_w$ | Density of water (mg/L) |
| $\tau$ | Dimensionless time |

| | |
|---|---|
| $\varphi$ | Reaction/diffusion Thiele modulus for intra-particle aqueous phase |
| $\varphi'$ | Reaction/diffusion Thiele modulus for external bulk fluid |
| $\varphi_{sm}$ | Reaction/diffusion Thiele modulus for soluble matter |
| $\omega_j$ | Number of particles in the jth particle size class |

# Executive Summary

Within the wastewater treatment industry there is a trend toward increasingly stringent discharge standards. Much of the increased scrutiny results from concerns that discharging treated wastewater into sensitive water bodies impacts downstream users and sensitive ecosystems. Additionally, water reuse is becoming more prevalent and highly treated reclaimed effluents are needed. The presence of microorganisms, including disease-causing pathogens, in wastewater treatment systems is a concern. In particular, effluent wastewater particles can harbor microorganisms and shield the occluded microorganisms from disinfectants such as chlorine. To ensure that the presence of occluded microorganisms in treated effluents is sufficiently reduced, particles must either be completely removed using filtration or sufficiently disinfected such that the occluded microorganisms are inactivated. Optimizing filtration and chlorine disinfection systems to reduce the concentration of viable occluded microorganisms in wastewater effluents is a valuable practice for the treatment industry and was the focus of this research.

## ES.1 Use of This Document for Evaluating Filtration and Disinfection Systems

There are two major outcomes presented in this report:

- A methodology for implementing site-specific optimization protocols.
- A summary of results for optimization protocols that were conducted at seven facilities.

The purpose of the optimization methodology is to provide design engineers and managers of treatment systems with tools that they can use to evaluate the impact of particles in their disinfection systems. The benefit of using the optimization protocols is that facility managers can identify treatment system conditions (particle counts and chlorine dose) that yield acceptable results for the reduction of occluded target microorganisms in their effluents. An example of how the optimization protocols were applied at one facility is detailed in Appendix D. The purpose of applying the optimization protocols at seven facilities was to demonstrate the application of the approach and to produce data that are representative of treatment facilities using different treatment trains and operating conditions.

Managers of treatment facilities have the option to: 1) Review the results obtained for facilities with similar operating conditions, or 2) Follow the step-by-step procedure to ensure adequate removal and inactivation of occluded microorganisms in particles. First, managers of treatment facilities can identify facilities that were used in this study that have similar characteristics to their own facility (e.g., using MCRT) and compare the results with their own operating conditions (e.g., initial chlorine concentration in the disinfection system and chlorine contact time). A summary of the results from this study is shown in Figure ES-1, which summarizes the combinations of particle counts (for particles with diameters larger than 11 μm) and initial chlorine concentrations that resulted in total coliform counts of 2 MPN/100 mL. A chlorine contact time of 90-minutes was used to obtain the results. The guideline of 2 MPN/100 mL was selected because it is consistent with existing regulatory requirements, and it represents the detection limit for the total coliform assay.

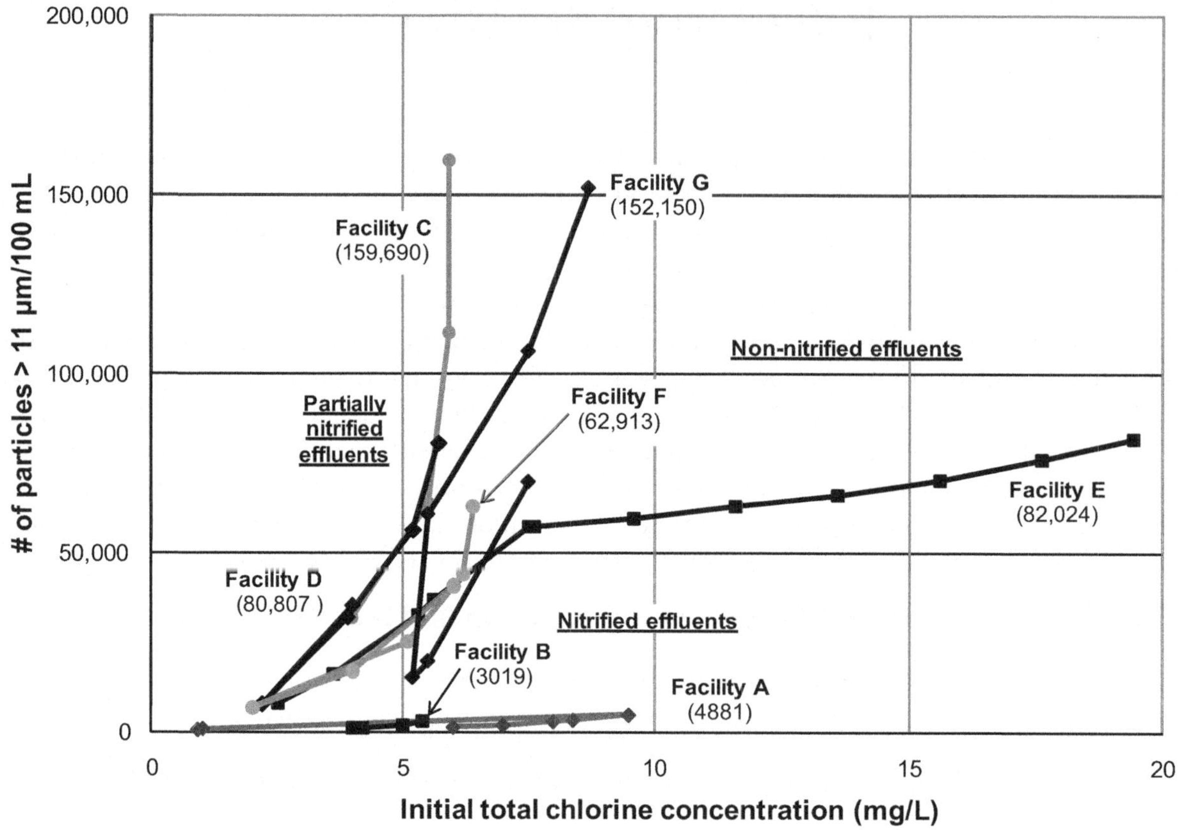


**Figure ES-1. Combinations of Particle and Initial Total Chlorine Concentrations That Resulted In Occluded Total Coliform Concentrations of 2 MPN/100 mL, as Predicted Using Model Output (Original Particle Counts Per 100 mL for Each Facility Are Listed in Parentheses).**

Two of the facilities included in the study, Facility A and B, have nitrified effluents (ammonia nitrogen < 0.02 mg/L) and low particle counts. Data collected from both of these facilities yielded a relatively flat slope as shown in Figure ES-1. For Facilities A and B small reductions in particle counts would result in large decreases in the quantity of chlorine required to meet the treatment objective of 2 MPN/100 mL. Facilities C and D have partially-nitrified effluents with ammonia nitrogen concentrations of approximately 1 mg/L. The slopes of the data from Facilities C and D are much higher than the slopes of the data from Facilities A and B, indicating that reductions in particle counts would have less of an impact on the chlorine required and that there is a more narrow band of chlorine concentrations that would be required to reduce total coliform concentrations to 2 MPN/100 mL. Facilities E, F, and G have non-nitrified effluents with ammonia nitrogen concentration higher than 20 mg/L. Interestingly, the slope of the data from Facility E is relatively flat and more similar to the slopes of the lines for the nitrified effluents from Facilities A and B although the particle count for Facility E was much higher than what was observed for Facilities A and B. The slopes of the data from Facilities F and G are high, similar to the slopes of the data for the partially-nitrified effluents from Facilities C and D. The particle counts for the partially-nitrified and non-nitrified effluents are much higher than those for the nitrified effluents. The information in Figure ES-1 demonstrates trends between operating conditions (e.g., ammonia nitrogen concentration) and the relationship between particles and disinfection system performance, although site-specific variability is also apparent.

With information regarding chlorine dose and particle counts, facility managers can potentially plot their facility's data on Figure ES-1 and make inferences about the effects that changes in the particle counts (achieved using filtration) would have on downstream chlorine disinfection systems. If a facility manager would like to perform a site-specific analysis and obtain more accurate estimates of their disinfection system performance, implementation of the optimization protocols is recommended. A step-by-step overview of the protocols is contained in Section 1-9, with references to the relevant chapters of this report. Figure ES-2 contains a schematic that summarizes the major components of the optimization protocols and references the relevant chapters in this report where readers can find more detailed information.

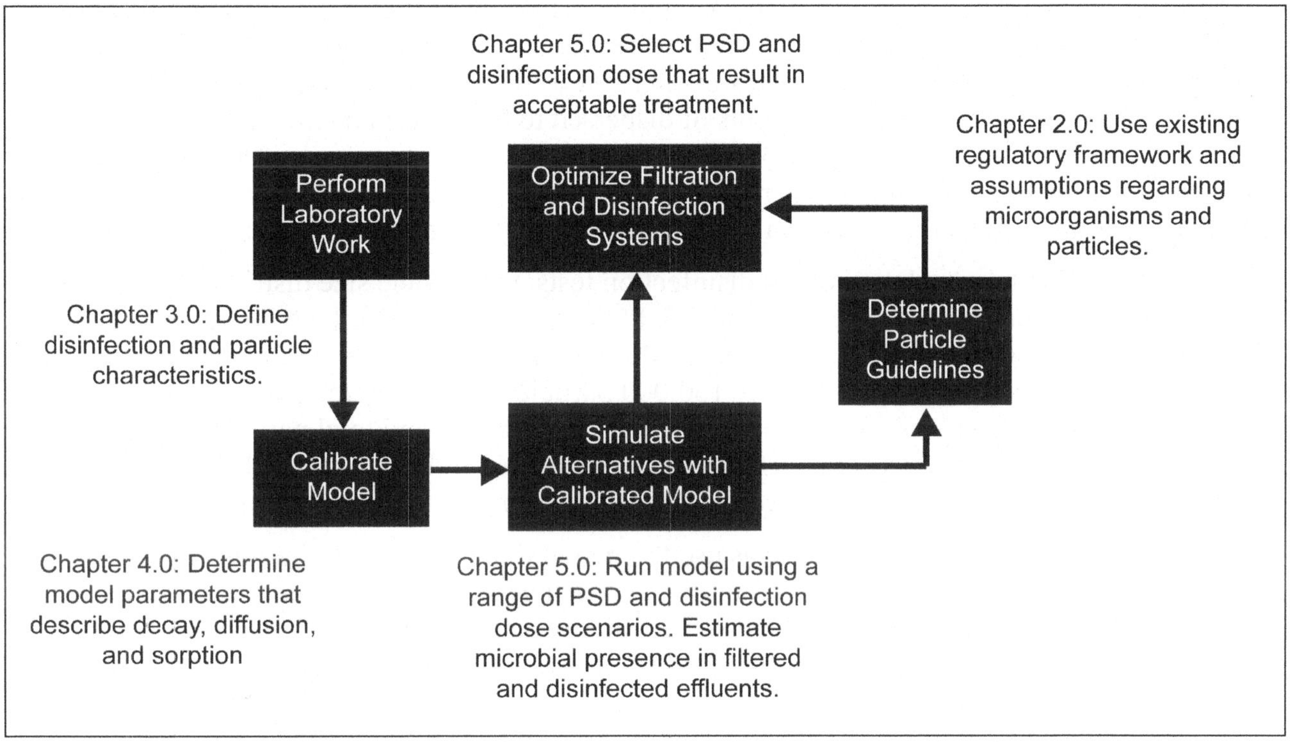


**Figure ES-2. Optimization Protocols for Reduction of Occluded Target Microorganisms in Filtration and Disinfection Systems.**

As shown in Figure ES-2, the sequence of tasks in the optimization approach includes collection of laboratory data, calibration of a computer model, collection of model output in simulations, concurrent development of particle guidelines, and identification of treatment conditions that result in acceptable reduction of occluded microorganisms. Laboratory data are needed to run and calibrate the model. Once the model is calibrated, it can be run with different assumptions regarding particles and the initial chlorine concentration. Model output consists of intra-particle chlorine concentrations that are spatially distributed throughout particles and are calculated for various time steps in a disinfection test. Using the intra-particle chlorine concentrations it is possible to calculate concentrations of occluded target microorganisms, which are then compared to a treatment objective. Concurrent development of particle guidelines is intended to provide facility managers with treatment metrics that they can use to determine if particles concentrations are sufficiently low to reduce the concentration of occluded microorganisms or if disinfection systems must be operated to inactivate occluded microorganisms.

## ES.2 Implementation of Optimization Protocols at Seven Facilities

Optimization protocols were implemented at seven facilities. Five of the plants utilize a conventional activated sludge treatment strategy, one facility uses oxidation ditches, and one facility uses a pure-oxygen activated sludge treatment system. As expected, given the range of treatment processes used, the water quality data, particle counts, and disinfection characteristics varied widely across facilities. As such, the most appropriate combinations for particle count reduction and chlorine concentration were unique to each facility.

Secondary effluent samples were collected from each facility and used in a series of tests to define the particle and disinfection characteristics. Turbidity, COD, pH, and ammonia nitrogen were measured to characterize the water quality conditions. Disinfection tests were performed to generate chlorine residual and total coliform data that were needed for model calibration. To determine the inactivation rate coefficients of dispersed total coliform bacteria, disinfection tests were conducted on filtered (2 μm) secondary effluent samples. Experiments were also conducted to collect information regarding the percentage of particles containing total coliform bacteria and the density of total coliform bacteria in the subset of particles containing total coliform bacteria.

In addition to completion of the disinfection tests, the particle size distribution (PSD) of each sample was analyzed for use in the model. PSD data collected for the facilities revealed large concentrations of particles in secondary effluent samples. The highest total particle count observed in this study was approximately 160,000 particles with diameters larger than 11 μm per 100 mL. Although there was a trend between measured turbidity and total particle counts, higher particle counts did not always coincide with higher measured turbidity values. Particle wet densities were also measured for the secondary effluent samples to provide data for the model.

A radial pore diffusion model was used to model the mass transfer of chlorine from the bulk fluid into spherical particles with four distinct particle sizes. Model output consisted of intra-particle chlorine concentrations that were spatially distributed throughout the particles; the concentrations are calculated for eight time steps following the introduction of chlorine in a disinfection test. The following data were used to run the model: PSD, initial chlorine concentration, and the particle volumetric void ratio (calculated using the measured particle wet densities). Chlorine residual concentrations observed during the disinfection tests were used to calibrate the model. Total coliform bacteria concentrations observed in the tailing portion of the inactivation curves were also used for model calibration.

The model was calibrated independently for each of the seven facilities. The selection of model parameters varied according to the operating parameters and water quality conditions at each facility. Once the model was calibrated, model simulations were run with different assumptions regarding particle counts and initial chlorine concentrations. The intra-particle chlorine concentrations generated in the model simulations were used to calculate estimated concentrations of occluded total coliform bacteria surviving the disinfection period. The final step of the optimization protocols was to compare the calculated total coliform bacteria concentrations with a treatment goal and determine combinations of particle and initial chlorine concentrations that yielded acceptable concentrations of occluded total coliform bacteria in the effluent. The treatment goal used in this study was an effluent limit for total coliform bacteria of 2 MPN/100 mL, which is consistent with existing regulatory guidelines. The conditions at the facilities that resulted in meeting the treatment goal of 2 MPN/100 mL are indicated in Figure ES-1. The optimization protocols are helpful to managers of treatment facilities because the results elicit the relationship between particles and the impact of performance on downstream disinfection systems.

As a model simulation exercise, additional experimental data were collected using three samples from Facility A collected on different days. Multiple disinfection tests were performed on each day and for some of the tests the sample was first filtered prior to use in the disinfection to remove the largest particles. The additional data were used for two purposes. First, the data were used in combination with the model parameters determined from the original calibration exercise to determine the effectiveness of the calibrated model for data collected on different days from the same facility. Second, the data was used to independently calibrate the model to determine the range of model parameters that result when multiple data sets are collected from the same facility. The model was most appropriately applied when calibration was performed independently for each sample; however, use of the calibrated model for obtaining information about intra-particle chlorine concentrations for various samples proved useful for obtaining a range of possible conditions.

## ES.3 Particle Guidelines

In this study the concept of particle guidelines is introduced. Particle guidelines represent the maximum concentration of particles (with diameters greater than 11 μm) that can be present in an effluent stream, which ensure that the concentration of occluded target microorganisms is below the microbial limit that formed the basis of the particle guideline. Particle guidelines are a function of exposure scenarios, microbial limits, and assumptions regarding the presence of microorganisms in particles. Microbial limits that form the bases of particle guidelines were adopted from existing regulatory guidelines. The limits correspond to treatment goals that were used to evaluate the total coliform concentrations calculated as part of the optimization protocols. Particle guidelines only ensure the reduction of occluded target microorganisms; disinfection systems must be in place and effectively operated to inactivate the dispersed fraction of microorganisms. If a facility cannot meet the particle guidelines, the disinfection system must be designed and operated to ensure that intra-particle disinfectant conditions are such that the occluded target microorganisms are sufficiently inactivated.

Two variables that describe the presence of microorganisms in particles and are needed to calculate particle guidelines are: 1) The frequency of association of microorganisms in particles (F), and 2) The microbial density of microorganisms in the subset of particles containing microorganisms (P). Experiments were conducted to measure F and P. The results for F were lower than expected when compared with published values, which were obtained using other methods. In lieu of using experimental data for F, a more conservative approach was taken and a value of one was assigned to F. Values of one were also assigned to P. To ensure that the approach was conservative, occluded microorganisms were assumed to be present at the center of particles, such that the entire particle would need to be sufficiently saturated with chlorine to inactivate all occluded microorganisms regardless of the concentration of microorganisms inside of the particles.

## ES.4 Application of the Optimization Protocols

The approach presented herein for ensuring adequate disinfection of microorganisms occluded in particles is unique. Ideally pathogens would be monitored directly in wastewater effluents to ensure their absence prior to use in water reuse projects or discharge into sensitive water bodies. However, pathogens are not easily monitored and indicator organisms are typically assayed in lieu of pathogens. The time-consuming nature of microbiological tests makes it difficult to quickly adjust treatment processes in response to increased concentrations of microorganisms. Use of the radial pore diffusion model makes it possible to estimate the effects of varying particle counts and initial chlorine concentrations. Once the model is calibrated using site-specific data collected in experimental studies, it is possible to run simulations and estimate the survival of microorganisms potentially present in particles. Although the optimization protocols were used in this research to study the inactivation of occluded total coliform bacteria using free and combined chlorine disinfectants, it is possible to conduct similar studies using other target microorganisms and chemical oxidants.

Both filtration and disinfection are used to reduce the concentration of microorganisms occluded in effluent particles. However, even the best filtration system cannot completely eliminate large (diameter greater than 11 μm) particles from the effluent, which are a result of imperfect filtration and the potential for amalgamation of smaller particles into larger particles post-filtration. Disinfection systems must be designed to sufficiently penetrate the interiors of residual particles. The intended benefit of this research is that managers of treatment facilities can make informed decisions regarding their filtration and disinfection systems and adequately reduce the concentration of microorganisms in effluent particles.

## ES.5 Potential to Extend the Current Approach to Pathogens

The optimization protocols were developed based on inactivation of total coliform bacteria in disinfection studies since total coliform bacteria are frequently used in regulatory policies. Application of the optimization protocols to pathogens and development of particle guidelines with a pathogen basis is preferred, but it would be difficult to perform the optimization protocols described herein using pathogens as the target microorganisms. Analytical methods for pathogens are specialized, labor-intensive, time-consuming, and costly. Instead of directly monitoring pathogens in disinfection tests, intra-particle chlorine concentrations obtained from the model that was calibrated with total coliform data were used to estimate the survival of pathogens potentially present in particles. To estimate the inactivation of occluded pathogens, inactivation rate coefficients from the literature were used. A disadvantage of using published inactivation rate coefficients was that the rates were not reflective of the water matrices being studied. Also, disinfection data were typically obtained for lab strains of pathogens that were suspended in saline solutions; disinfection with combined chlorine was not always considered.

Use of the optimization approach to ensure that occluded pathogens in effluent particles are adequately reduced is a worthwhile goal. Intra-particle chlorine concentrations that are high enough to inactivate occluded total coliform bacteria may not be high enough to inactivate occluded pathogens. Although the results of this study were dependent on the pathogen of interest and the treatment goal, sufficient reduction of potentially occluded pathogens was not achieved for most of the treatment scenarios considered. The results highlight the importance of considering pathogens potentially present in particles, and of conducting site-specific analyses.

Calculation of pathogen-based particle guidelines is possible although significant assumptions are necessary due to the paucity of current data. Development of pathogen treatment goals was performed using risk assessment methods and existing regulatory guidelines for treatment in combination with published data of the presence of pathogens in source waters (e.g., log removal in treatment processes). Data regarding the presence of pathogens in particles are also limited and the same assumptions made for total coliform bacteria were applied to pathogens. The resulting particle guidelines were very low, reflecting particle concentrations that are lower than what can be obtained using current filtration systems, especially given the potential for re-formation of particles in post-filtration effluents. The inability of facilities to meet the particle guidelines suggests that disinfection systems must be designed such that the intra-particle chlorine concentrations are sufficient to inactivate pathogens occluded in particles.

## ES.6 Future Direction for Optimization Protocols

The optimization protocols are beneficial to managers of treatment systems. In the future, it would be ideal to repeat the protocols at a single facility to better define the range of filtration and disinfection system characteristics, including characterization of diurnal and seasonal trends and responses to plant upsets. Also, the reliability of the optimization protocols would be higher if some of the underlying assumptions were verified, including the values of F and P and the assumption that occluded microorganisms have the same inactivation characteristics as dispersed microorganisms. Ultimately, the health risks related to particles in treated effluents need to be investigated such that treatment goals can be established that are protective of public health.

# CHAPTER 1.0

# INTRODUCTION AND BACKGROUND INFORMATION

## 1.1 Impact of Particles in Wastewater Disinfection Systems

One of the reasons that viable microorganisms and infectious viruses can persist in treated wastewater is related to the presence of particles. Effluent wastewater particles are often residual activated sludge particles, which consist of densely packed bacterial cells, other microorganisms, miscellaneous debris, and organic matter. In addition to carryover from the secondary treatment process, effluent particles can be present due to microbial growth in plant pipelines and channels and due to the amalgamation of small particles.

Even highly treated wastewater effluents contain a large number of particles and these particles can cause a shielding effect, where microorganisms and viruses are protected from disinfectants as a result of their occlusion in particles (Berman et al., 1988; Hejkal et al., 1979; Herson et al., 1987; Madge and Jensen, 2006; Ormeci and Linden, 2002). Chemical oxidants and UV light must first penetrate through particle pores and protective organic materials to inactivate occluded microorganisms and viruses. If the penetration of disinfectants into particles is incomplete, not all of the particle-associated microorganisms are inactivated. Diffusion of disinfectants through particles is largely reliant on particle's properties, including size. In the case of UV disinfection, particles that have diameters larger than 11 μm contain regions that are not exposed to the UV light applied in the bulk liquid medium regardless of increased UV intensity and exposure time (Emerick et al., 1999). To produce treated effluents that are reliably suitable for beneficial use, a combination of filtration and disinfection is often necessary (Braunstein et al., 1996; Jolis et al., 1999).

In disinfection systems tailing is observed where a stable residual concentration of microorganisms persists despite high concentrations and/or contact times of the disinfectant. Where tailing is present, inactivation of the microorganisms proceeds slowly with increased application of disinfectant dose. Tailing can be caused by more than one phenomenon. For example, tailing can be caused by the variability of resistance that the target microorganisms have to the applied disinfectant. There may be a subpopulation of microorganisms that is more resistant to the disinfectant, and therefore, must be exposed to a higher amount of the disinfectant before inactivation can occur (Ridgway and Olson, 1982). An additional tailing phenomenon occurs when the disinfectant concentration is decreasing during the disinfection period. For example, the microorganisms themselves may be reducing, or quenching, the disinfectant concentration during the contact time and causing reduced inactivation (Lambert and Johnston, 2000). Tailing can also occur as a result of the presence of microorganisms in particles (Berman et al., 1988) or as a result of organic matter adhering to and protecting microorganisms (Cantwell et al., 2008). Tailing can occur in disinfection systems that rely on chemical oxidants if the concentration of the disinfectant is insufficient, such that it reaction of the chemical oxidant is complete prior to complete inactivation of target microorganisms.

Where the residual concentration of microorganisms is lower than permit limits, tailing is not a problem. However, as is often the case for facilities with stringent permit limits, if the residual concentration of microorganisms is higher than the permit limits, changes must be made to the operational strategy. To sufficiently reduce the concentration of microorganisms in wastewater effluent particles, either the particles must be removed using filtration or the disinfection system must be designed and operated to ensure that the occluded microorganisms are inactivated. Ideally, filtration and disinfection systems would operate in tandem to reduce effluent microbial concentrations. For example, it may be more feasible to remove large particles using filtration and to operate disinfection systems to penetrate all regions of smaller particles.

## 1.2 Previous Work Regarding Disinfection of Microorganisms Occluded in Particles

Several approaches have been used to study the disinfection of microorganisms occluded in particles. Madge and Jensen (2006) serially filtered wastewater using nylon filters; the particles that collected on the filters were washed with a phosphate buffer into filtered (0.45 μm) wastewater solutions and resuspended. The technique of Madge and Jensen (2006) is intended to introduce particles containing microorganisms into solution in the absence of dispersed organisms. Use of filtered wastewater for the bulk solution ensures that the bulk chemical aqueous characteristics are maintained. Berman et al. (1988) used a similar technique of isolating wastewater particles into different size classifications using filtration and resuspending the filtered particles in buffered solutions. In lieu of studying indigenous organisms, some researchers choose to introduce laboratory strains into samples. For example, Templeton et al. (2005) added phages to synthetic water samples along with activated sludge particles and mixed the resulting solution in a jar test apparatus to encourage particle association of the phages. The advantage of the technique used by Templeton et al. (2005) is that the total quantity of phages can be controlled as well as the water quality of the stock solution.

A disadvantage of isolating particles on filters and re-introducing them into solution is that the particles could potentially be altered during the laboratory procedures, thus changing the viability of occluded microorganisms. The particle size distribution and organic matter composition may also be altered. A disadvantage to introducing cultured microorganisms is that the introduced microorganisms may not associate with particles at the same frequency as originally occluded microorganisms and they may not be dispersed throughout the particles in the same manner (e.g. there may be preferential association with the outer regions and surface of the particles). In this work the researchers aimed to replicate operating conditions at a full-scale disinfection system as much as possible, and chose not to concentrate particles and to use indigenous microorganisms as the target. In some cases they chose to filter samples in the laboratory to observe the dispersed microorganisms or to observe disinfection characteristics in the absence of larger particles.

## 1.3 Particle Guideline Approach

In this work the researchers propose the use of particle guidelines for treatment plant effluents that are more informative than traditional water quality metrics such as turbidity. Particle guidelines are calculated particles concentrations that are sufficiently low to ensure that the discharge limits for the target microorganism are met, assuming that the disinfection system is sufficiently designed and operated to inactivate the dispersed target microorganisms. No inactivation of occluded target microorganisms is assumed when calculating the particle guidelines. If the particle guidelines cannot be met at a particular facility, the disinfection system process control parameters (e.g. chlorine concentration and contact time) must be applied such that at least a fraction of the occluded target microorganisms is inactivated. It may be difficult for many facilities to rely on their filtration systems to meet the particle guidelines; a combination of filtration and disinfection is often necessary to sufficiently reduce the concentration of target microorganisms. The particle guidelines were calculated based on existing regulatory standards for indicator organisms, which are based on historic implications for the effect of regulations on human health. In future efforts, when the proposed protocols are extended to address pathogens occluded in particles, the researchers advocate using risk-based standards that are formulated using a risk assessment framework.

Disinfection studies are typically based on a single microorganism or on a relative small number. However, at any given time wastewater effluents can contain a large number and diversity of species. To account for the variety of microorganisms potentially present in wastewater, the researchers propose development of guidelines based on a variety of microorganisms and finally, implementation of the most conservative guidelines. The ultimate goal of this research is to ensure that intra-particle chlorine concentrations are sufficiently high to inactivate occluded microorganisms.

## 1.4 Indicator Organisms and Pathogens

Pathogenic microorganisms are often not directly measured in wastewater effluents as a result of their intermittent presence and low concentrations. Pathogen assays are costly and time-consuming, requiring specialized facilities and personnel. Indicator organisms, such as total and fecal coliform bacteria, are often used as surrogates for the removal of pathogens in treatment processes. A significant difference between monitoring for pathogens and indicator organisms is that pathogen testing requires a much larger sample volume as a result of the lower concentration of microorganisms present. Detection limits for pathogen tests are dependent on the sample volume analyzed. Testing for pathogens in highly treated wastewater can result in a large percentage of negative results. Increasing the sample volume and increasing the number of tests can elicit more information regarding the presence of pathogens. There are studies available where pathogens were observed in wastewater influents, effluents, and throughout the treatment process (e.g. Rose et al., 1996; Rose et al., 2001; and Rose et al., 2004). In Florida, *Giardia* and *Cryptosporidium* testing is required for water reuse projects although many of the samples contain concentrations that are below detection limits (York et al., 2003). There are many other studies of pathogens in wastewater influents, effluents, and throughout the treatment processes; however, it is outside the scope of this project to identify all of these studies.

Effluent wastewater particles are composed primarily of bacterial cells and associated organic matter; the bacteria present represent a biological diversity of bacterial species (Bitton 2005; Jenkins et al., 2003). Despite the diversity of microbial species in wastewater, scientists are constrained by the availability of microbiological methods that can detect and enumerate viable target microorganisms. For example, it is difficult to identify an appropriate indicator organism that has characteristics that are similar to those of pathogens. Most bacteria present in wastewater are not culturable, meaning that they cannot be isolated and grown in the laboratory using currently available growth media. The choices available for monitoring microorganisms in wastewater are further limited by the types of microorganisms that are reliably present in wastewater systems. For example, some indicator organisms are reliably present in untreated wastewater, but not in treated effluents.

In this work indicator organisms (total coliform bacteria) were monitored in experimental disinfection studies to calibrate a model and develop optimization protocols for treatment facilities. As such, the optimization approach described herein is based on the presence and persistence of indicator organisms. An advantage of this approach is that current regulatory guidelines are based on indicator organisms, which are often coliform bacteria. Also, using indicator organisms makes it possible for utilities to follow this approach since they routinely monitor indicator organisms and may not have the resources available to monitor pathogens. If additional data become available for pathogens in the future, it would be possible to extend the current optimization approach to pathogens and develop a risk-based approach that is based on pathogens.

## 1.5 Tertiary Filtration Systems

Tertiary filtration typically reduces the total number of particles and alters the shape of the particle size distribution (PSD) curve. Filtration is accomplished using a variety of technologies; the resulting effluent particle concentrations and size distributions are dependent on the type of technology employed (e.g., microfiltration or gravity filtration). One example of how filtration alters the PSD is that following filtration the percentage of large particles often decreases such that the proportion of small particles increases. In filtration systems, the relationship between microbial presence in particles and particle size determines the extent of microbial removal. In this study, information regarding particle removal rates in filtration systems was necessary to identify a range of reasonable PSD values for post-filtration effluents. If facility managers find that the PSD values necessary to meet particle guidelines are unobtainable using the filtration technologies in place, it may be necessary for them to identify alternative filtration technologies that are capable of producing acceptable effluents. Alternatively, facility managers may opt to alter the operating conditions of their disinfection systems and disinfect occluded microorganisms inside of removing them.

There are limited data on the impact of filtration on PSD, and all of the available data indicate the site-specific nature of filtration system performance. Kobler and Boller (1997) collected PSD data for seven filtration systems in Switzerland, representing four different filtration technologies. In most cases removal rates greater than 80% were observed for particles larger than 10 µm. However, in some cases removal rates were poor (less than 50%) for larger particles. Kobler and Boller (1997) also investigated the effect of flocculation on PSD for five treatment systems. In one case the dose of flocculant was small and there was no change in the PSD. In three of the treatment systems inappropriately designed flocculation systems resulted in incomplete particle formation and there was an increase in small particles without a substantial increase in large particles. In one case this problem was overcome by contact filtration, or additional flocculation in the filter bed. The observations of Kobler and Boller (1997) make it clear that the PSD of a filtration system should be monitored to ensure that adequate removal rates are obtained for various particles sizes.

In a pilot study, van der Graaf et al. (2001) found that introduction of a flocculant had a negative impact on filtration in terms of reducing the number of effluent particles with diameters larger than 10 µm. Surprisingly, van der Graaf et al. (2001) observed a reduction in turbidity despite the increased total volume of particles, which may have resulted from modest (less than 20%) reductions in the number of particle with diameters less than 10 µm. In another pilot study, Kaminski et al. (1997) found that flocculants had a positive impact on filtration for removal of particles, but that the filtration rate and backwashing frequency must be adequately controlled.

Kaminski et al. (1997) observed varying changes in the PSD that were dependent on filter grain size, filtration rate, and filter run time when the samples were collected. During the middle of the filter run, the filters removed approximately 10-40% of particles in the 10-15 μm size range while approximately 80-90% of particles in the 100-150 μm size range were removed (Kiminski et al., 1997).

## 1.6 Chlorine Disinfection Systems

Chlorine is still the most commonly used disinfectant for water and wastewater treatment in the U.S. Although UV is also a popular disinfectant, chlorine disinfection is the focus of this study. Chlorine is typically added as chlorine gas, sodium hypochlorite (NaOCl), or calcium hypochlorite ($Ca(OCL)_2$) (Tchobanoglous et al., 2003). In the absence of ammonia, the chlorine introduced into the water stream will form hypochlorous acid (HOCl) and hypochlorite ($OCl^-$). Below a pH of two, aqueous chlorine in the elemental form ($Cl_2$) can also be present, but this pH is outside of the range of typical treatment systems. The three species HOCl, $OCl^-$, and $Cl_2$ constitute the free chlorine residual. The percentage of the chlorine products formed following chlorine addition depends on both the system pH and temperature. The partitioning of the free chlorine species is important because HOCl is a much more effective disinfectant than $OCl^-$ although the specific efficiency of disinfection will depend on the species disinfected (Snoeyink and Jenkins, 1980). If the pH is between approximately 6.0 and 8.5 a combination of HOCl and $OCl^-$ will be present.

In the presence of ammonia, combined chlorine species will form, including monochloramine ($NH_2Cl$) and dichloramine ($NHCl_2$). The combination of the combined species and the free chlorine residual constitutes the total chlorine residual. When chlorine is added to water that contains ammonia, the chlorine will form combined chlorine species until the ammonia demand is met and then any additional chlorine added will result in a free chlorine residual; this is referred to as breakpoint chlorination.

The basic concepts of chlorine chemistry have several repercussions for the research proposed herein. The pH dependence of the speciation of free chlorine residuals between pH values of 6.0 and 8.5 dictate the need for pH control during the disinfection tests. When NaOCl is added to water, hydroxide ions are formed and the pH rises. The rise in pH can be counteracted by adding an acid such as hydrochloric acid (HCl). Likewise, changes in temperature should be avoided throughout a disinfection test to ensure consistent results.

The differentiation between disinfection for nitrified and non-nitrified effluents is important because microorganisms respond differently to combined and free chlorine products. Typically, combined chlorine reacts and disinfects much more slowly than free chlorine. As such, nitrified and non-nitrified effluents may have different regulatory constraints. In nitrified effluents, the chlorine is present as free chlorine and the total chlorine residual is nearly equivalent to the free chlorine residual. In non-nitrified effluents the chlorine added is present as combined chlorine products as long as breakpoint conditions have not been reached. For partially nitrified effluents, a combination of combined and free chlorine products is present.

The concentration of effluent ammonia in wastewater treatment plant effluents is dependent on the type of secondary treatment used, and is typically a function of the mean cell residence time (MCRT). Facilities that have a short MCRT typically have higher effluent ammonia concentrations. A long MCRT is necessary to accommodate the slow growth of nitrifying bacteria that facilitate the oxidation of ammonia to nitrite, as well as the conversion of nitrite to nitrate. An operating rule of thumb that has dominated chlorine disinfection of wastewater and is used by the state of California for regulation of reclaimed water is a CT (chlorine residual concentration multiplied by contact time) of 450 mg*min/L (CA DPH, 2001). This CT guideline was based on a study completed in 1968 and does not address the site-specific nature of disinfection; however, a CT of 450 mg*min/L is thought to provide 4-log removal of poliovirus (Asano et al., 2007, p. 617). In addition, chlorine disinfection is also influenced by organic matter, metals, and industrial components in the wastewater including surfactants that can greatly increase chlorine demand.

## 1.7 Disinfection Kinetics

Inactivation of microorganisms involves inactivation of both the dispersed and occluded fractions. In this study, first-order kinetics were assumed and identical inactivation rate coefficients were used dispersed and occluded microorganisms. Dispersed microorganisms were defined as microorganisms that pass through a 2 μm filter. The inactivation rate of occluded microorganisms was assumed identical to that of dispersed organisms; however, inactivation of occluded microorganisms was assumed delayed the diffusion of chlorine ions that must penetrate particle pore spaces and obstructing organic materials before inactivating occluded microorganisms. Equation 1-1 describes inactivation in systems where there are dispersed and occluded populations of microorganisms (Dietrich et al., 2007). Equation 1-1 is based on inactivating occluded microorganisms that are located in the center of particles, so that particles must have a sufficiently high dose of chlorine at the particle center before particles containing microorganisms no longer contain microorganisms.

$$N(t) = N_d e^{-k_i \int_0^t c_{bulk}(t)dt} + \sum_j F_j \omega_j e^{-k_i \int_0^t c_i(r=0,t)dt} \qquad (1\text{-}1)$$

In Equation 1-1, *N(t)* is the concentration of microorganisms following exposure to a disinfectant for time t (microbial units/100 mL), $N_d$ is the initial concentration of dispersed microorganisms (microbial units/100 mL), $k_i$ is the inactivation rate coefficient (L/mg/min), $c_{bulk}$(t) is the residual chlorine concentration (either free or total chlorine residual) in the bulk fluid at a given time (mg/L), *t* is the contact time (min), $F_j$ is the fraction of particles in the jth particle class that contain microorganisms, $\omega_j$ is the concentration of particles in the jth particle class, and $c_j$(r=0,t) is the residual chlorine concentration (either free or total chlorine residual) at the particle center at a given time for particles in the jth particle class (mg/L).

The inactivation rate coefficient ($k_i$) was determined using data from the disinfection tests conducted on filtered samples (Equation 1-1 with only the first term on the right hand side of the equation), and was used in combination with intra-particle chlorine concentrations calculated with the radial pore diffusion model to determine the survival of microorganisms following exposure to chlorine. The concentration of dispersed microorganisms is typically several orders of magnitude higher than the concentration of occluded microorganisms such that the initial concentration of dispersed microorganisms is essentially the same as the total concentration of microorganisms at the beginning of a disinfection test. The concentration of occluded microorganisms is identified by extending a best-fit line, obtained using data from the tailing region of the inactivation curve, through the y-axis to determine the microbial concentration when the dose or time (depending on how the curve is plotted) is zero.

In the absence of particles, Equation 1-1 can be written as Equation 1-2, which is the equation that was used to determine the inactivation rate coefficient of dispersed microorganisms using data from disinfection tests performed on filtered samples. Equation 1-2 was also used to estimate inactivation of microorganisms occluded in particles using intra-particle chlorine concentrations calculated by the model.

$$N(t) = N_d e^{-k_i \int_0^t c_{bulk}(t)dt} \qquad (1\text{-}2)$$

## 1.8 Project Objectives

The principal goal of this research was to develop protocols for optimizing filtration and chlorine disinfection systems to reduce the concentration of microorganisms occluded in effluent wastewater particles. The optimization protocols were intended for implementation by managers of treatment facilities to better understand the role of particles in their disinfection systems and to make better decisions regarding design and operation of filtration and disinfection systems. Concurrent with these efforts, particle guidelines were developed for comparison with observed particle counts, making it possible to determine if filtration alone was sufficient to remove occluded target microorganisms. Where filtration was not sufficient, a radial pore diffusion model was run to estimate intra-particle chlorine concentrations, and subsequent calculations were performed to estimate the inactivation of occluded target microorganisms for a variety of initial chlorine and particle concentrations. The final product was a summary of combinations of initial chlorine and particle concentrations that resulted in concentrations of occluded target microorganisms that met regulatory limits.

In this study, the optimization protocols were performed at seven facilities and were repeated three times at one facility to evaluate the range of values selected for model parameters. Samples collected from all seven facilities were subjected to a series of laboratory tests to quantify key disinfection and particle characteristics. Laboratory data were used to calibrate a radial pore diffusion model and estimate the intra-particle chlorine concentrations at various radial distances within the particles and at various time steps. The calibrated model was used to run simulations with varying initial chlorine and particle concentrations. The resulting intra-particle chlorine concentrations were used in combination with other data to determine the concentrations of effluent occluded microorganisms for different particle count and chlorine concentration scenarios. The final product of this research was information that relates treatment plant conditions to estimates of inactivation of occluded microorganisms for seven facilities as well as optimization protocols that could be completed at any given facility. A step-by-step summary of the optimization protocols is contained in the following section.

## 1.9 Step-By-Step Project Approach

The project described herein is intended to assist utility managers and treatment plant designers in the co-optimization for the design and operation of filtration and disinfection systems that ensure adequate inactivation of microorganisms occluded in wastewater particles. An example of how the approach was applied to one facility is detailed in Appendix D, which includes descriptions of the data collected and of the steps taken to calibrate the model. To apply the approach described herein, the following steps are necessary:

1) Collect a secondary effluent sample that has not undergone disinfection and perform the following laboratory procedures (Chapter 3.0):
   a) Filter a subset of the sample to remove particles large enough to harbor bacteria (e.g. with 2 μm filters).

b) Use the secondary effluent sample and the filtered sample to perform separate chlorine disinfection tests, collecting total coliform and chlorine residual data at regular time intervals.

c) From the data obtained from the disinfection test using a filtered water sample, calculate the first-order inactivation rate coefficient for total coliform bacteria not associated with particles.

d) Measure the particle wet density of the secondary effluent sample and use this information to determine the particle porosity.

e) Measure the particle size distribution (PSD) of the secondary effluent sample for particles larger than 11 μm. Where filtration is practiced, a filtered effluent sample should also be collected for PSD measurement to evaluate the effectiveness of filtration.

f) Subject a subset of the unfiltered secondary effluent sample to a low dosage of UV light to inactive total coliform bacteria not associated with particles and determine the total coliform concentration of the irradiated sample. Calculate the frequency at which total coliform are occluded in the particles by dividing the total coliform concentration by the total number of particles that are larger than 11 μm.

2) Using data from the laboratory studies, indentify information necessary to calibrate the radial pore diffusion model (Chapter 4.0):

a) Divide the PSD data into four particle size classes (each with a range of particle diameters) and determine the median for each particle size class.

b) Identify an appropriate diffusion coefficient using published values.

c) Identify values for the Biot number using data from published studies or make an assumption regarding appropriate Biot number values based on published information regarding typical ranges of values.

d) Calculate the number of dimensionless time steps necessary to run the model.

e) Determine the initial concentration of occluded total coliform bacteria by viewing the disinfection data and distribute the values across the four particle size classes, assuming that the occluded total coliform bacteria occupy larger particles prior to smaller particles.

f) Compose the necessary parameter estimating software (UCODE) and process model input files. The units of input information should be consistent (e.g. concentrations in terms of mg/L and particles/L and lengths in terms of dm, in this case the characteristic length is 1 dm because 1-L=1 $dm^3$).

3) Calibrate the radial pore diffusion model (Chapter 4.0):

a) Use the PSD data, particle porosity, and initial chlorine concentration in the disinfection test (performed without prior filtration) to run the radial pore diffusion model. The initial chlorine concentration used to run the model may be lower than what was actually added to the reactor in the disinfection test; a value should be selected that results in a best-fit of the model output, as determined in the first phase of model calibration. The initial chlorine concentration in the model is the concentration of chlorine added with the instantaneous chlorine concentration subtracted.

b) Use UCODE parameter estimating software to select optimal parameter values, which is accomplished using the observations of chlorine residual and total coliform concentrations. A better fit for the total coliform data may be obtained by log transforming the total coliform concentrations that are calculated by the model and that were observed in the experiments.

   c) Assign appropriate weights to the chlorine and total coliform observations. For example, weights can be assigned to the individual observations using an estimate of the variance or using the coefficient of variation. Weight multipliers can also be assigned to groups of observations to independently weight the chlorine and total coliform observations.

4) Run simulations using the calibrated model (Chapter 5.0).
   a) Alter the initial chlorine, particle and occluded total coliform concentrations in each simulation.
   b) Obtain values of the final occluded total coliform concentration from the model output.
5) Concurrent with the laboratory and modeling efforts, develop effluent particle guidelines using the existent regulatory framework (Chapter 2.0).
   a) Identify appropriate regulatory microbial limits (e.g. total coliform concentration limits of 2 MPN/100 mL).
   b) Establish guidelines based on assumptions regarding the presence of microorganisms in particles and on specific exposure routes (e.g. recreational contact with reclaimed water). Values can be assumed for the frequency of microbial association with particles (F) and for the density of microorganisms in particles containing microorganisms (P). Alternatively, determine values of F and P in experimental studies.
   c) Calculate particle guidelines using a variety of assumptions, and adopt the most conservative guidelines.
6) Compare the particle guidelines with observed particle concentration, and compare estimates of occluded total coliform presence (calculated by the model) with the corresponding treatment objectives (Chapter 5.0).
   a) Determine if the measured total particle counts are higher or lower than the particle guidelines.
   b) Identify combinations of initial chlorine and particle concentrations used in the model simulations that result in total coliform concentrations that are less than the treatment objective.
7) Extend the current protocols, to the extent possible, to address the presence of pathogens in particles (Chapter 6.0). <u>This step is optional.</u> At this time the use of optimization protocols and particle guidelines that are based on indicator organisms are preferred because data are currently lacking to extend the approach to that of a pathogen basis. However, in Chapter 6.0 an approach is presented that could be implemented in the future when additional data are available.
   a) Identify treatment goals for pathogens based on existing regulatory guidelines, risk assessment data, and existing data regarding the presence of pathogens in environmental waters needed to apply log-removal based standards.
   b) Similar to the particle guidelines that were developed in Chapter 2.0, develop particles guidelines based on pathogens. In the absence of data regarding the presence of pathogens in particles (F and P), make assumptions (e.g. F=1 and P=1). Calculate particle guidelines based on pathogens using a variety of assumptions

(various regulatory guidelines, organisms, exposure routes), and adopt the most conservative guidelines for a variety of exposure scenarios.

c) Compare the particle guidelines based on pathogens with observed particle concentrations to determine if treatment objectives for pathogens can be met using filtration alone, or if proper operation of the disinfection system is needed in inactivate occluded pathogens.

d) Run model simulations with different values for the inactivation rate coefficient that are representative of values measured for pathogens (available in the peer-reviewed publications). Estimate the effluent concentrations of occluded pathogens and compare these values with the treatment objectives for pathogens (e.g., $1 \times 10^{-9}$ PFU/100 mL).

# CHAPTER 2.0

# PARTICLE GUIDELINES

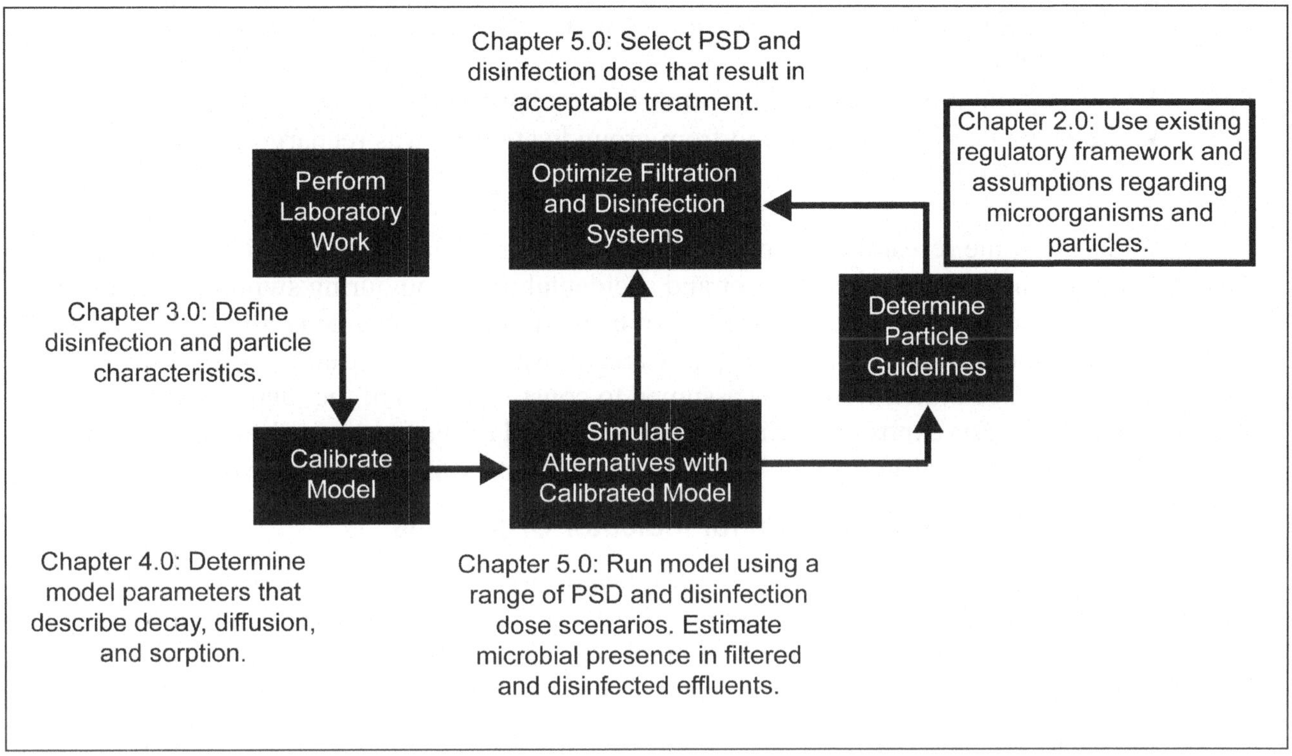


## 2.1 Introduction

Particle guidelines represent the maximum concentration of particles (with diameters greater than 11 μm) that can be present in an effluent stream, and ensure that the concentration of occluded target microorganisms is below the microbial limit that formed the basis of the particle guideline. Particle guidelines were developed based on the existing regulatory framework for indicator organisms that has historically demonstrated acceptable health risks. In addition to effluent limits for microorganisms, three additional constituents are needed to develop particle guidelines: the amount of water that an individual is exposed to, the frequency of occurrence of microorganisms within particles, and the microbial density within particles that contain microorganisms. Once particles guidelines are calculated, they can be used as a performance metric for particle count data collected from operational treatment facilities. Additionally, the particle guidelines can be compared with information gleaned from the radial pore diffusion model simulations to determine if operating treatment facilities would likely meet designated treatment objectives with various PSD and chlorine concentration scenarios.

## 2.2 Exposure Scenarios

Microbial standards for reclaimed water are necessary because of the potential impacts on downstream uses. The type and degree of exposure of humans to reclaimed water depends on the application. Examples of exposure routes include the following:

- Accidental ingestion at an irrigated site,
- Inhalation of aerosols at an irrigation site,
- Accidental ingestion during swimming in a water body where reclaimed water is discharged,
- Consumption of food crops irrigated with reclaimed water, and
- Consumption of water derived from groundwater that was recharged with a component of reclaimed water.

In this study the researchers considered two exposure routes: consumption of groundwater that includes reclaimed water and accidental ingestion during swimming in a water body that contains reclaimed water. Two liters of drinking water consumed daily was assumed, which is the arithmetic average value for total water consumption reported by Roseberry and Burmaster (1992). Reclaimed water was assumed to constitute 50% of the total water supply (Asano et al., 1992). For exposure during swimming, 50 ml of marine or fresh water consumed per hour of swimming with an average of 2.6 hours per swim was assumed (Haas et al., 1999).

## 2.3 Existing Microbial Standards for Indicator Organisms

The existing regulatory framework for indicator organisms typically relies on total coliforms, fecal coliforms, enterococci, and *Escherichia coli*. Regulatory effluent limits are based on the water quality of the receiving water body and the anticipated impact of the discharge. For water reuse projects, limits are a function of the intended use of the treated water. Effluent microbial limits in reclaimed water are determined based on the risk of human contact and the anticipated exposure.

For wastewater effluents discharged into recreational surface waters, the U.S. EPA has established the following risk criteria for swimming-related gastroenteritis resulting from recreational exposure: eight cases of illness occurring for every 1,000 swimmers at fresh water beaches, and 19 cases of illness for every 1,000 swimmers at marine beaches (U.S. EPA, 1986). The U.S. EPA criteria were developed as the result of epidemiological studies. The recreational risk criteria correspond to concentrations of indicator organisms shown in Table 2-1. When using the standards in Table 2-1, no fewer than five samples should be collected at intervals equally spaced over a 30-day period and the geometric mean of the data should be used.

**Table 2-1. Indicator Organism Counts per 100 mL Corresponding With Acceptable Risk for Wastewater Effluents Discharged into Surface Water Bodies Where There is Recreational Contact (U.S. EPA, 1986).**

| Organism | Fresh water beaches | Marine beaches |
|---|---|---|
| Fecal coliform | 200 | 200 |
| Enterocci | 33 | 35 |
| *E. coli* | 126 | -- |

In addition to the microbial limits shown in Table 2-1, the concentration of specified indicator organisms should not exceed one-sided confidence intervals that are calculated based on the frequency of bathing (U.S. EPA, 1986, p. 16). Calculation of log-standard deviations should be performed unless data are insufficient, and then use of specified values is acceptable (U.S. EPA, 1986, p. 16)

The California Recycled Water Criteria consists of several classifications for reclaimed water for which the potential beneficial uses are limited by the treatment standard that is met. The most stringent standard is for disinfected tertiary recycled water, which requires filtration with an approved technology (CA DPH, 2008) as well as a disinfection system meeting one of the following requirements: 1) Chlorine disinfection that is designed for a CT (chlorine residual concentration multiplied by the contact time) of at least 450 mg*min/L and a model contact time of at least 90-minutes, based on peak dry weather flow, or 2) An alternative disinfection process that can provide 5-log removal of viruses when used in combination with the filtration system (CA DPH, 2001, p. 46).

Additional reclaimed water standards exist for disinfected secondary-2.2 recycled water and disinfected secondary-23 recycled water, which require oxidation and disinfection but not filtration. To produce disinfected tertiary recycled water or disinfected secondary-2.2 recycled water, the median total coliform counts for a seven-day period must be less than 2.2 MPN/100 mL and there cannot be more than one total coliform count higher than 23 MPN/100 mL in a 30-day period (CA DPH, 2001, p. 45-6). For disinfected tertiary recycled water an additional requirement is that no sample for total coliform bacteria can be higher than 240 MPN/100 mL (CA DPH, 2001, p. 46). To produce disinfected secondary-23 recycled water, the median total coliform counts for a seven day period must be less than 23 MPN/100 mL and there cannot be more than one total coliform count higher than 240 MPN/100 mL in a 30-day period (CA DPH, 2001). Other states have adopted similar standards that correlate acceptable uses with microbial limits (Asano et al., 2007; U.S. EPA, 2004).

The National Drinking Water Standards dictate that all tests of total coliform bacteria, fecal coliform bacteria, and *E. coli* should yield results below detection limits (U.S. EPA, 2002). None of these microorganisms should ever be found in any samples. Although specific to drinking water, this standard for indicator organisms may be applied to reclaimed water projects where high standards of microbial water quality are warranted.

In summary, there are three guidelines for microbial limits that can serve as the bases for calculating particle guidelines with an indicator organism basis (Table 2-2). Guideline 1 was developed as a result of epidemiological studies and the other guidelines are based on analytical limits of the methods used and on the historic demonstration of minimal risk of using reclaimed water.

Table 2-2. Regulatory Guidelines for Indicator Organisms Used As the Bases for Particle Guidelines.

| | Source | Guideline |
|---|---|---|
| 1 | U.S. EPA, 1986 | Recreational water: fecal coliform=200/100 mL and fresh and marine beaches, enterococci=33/100 mL at fresh beaches, enterococci=35/100 mL at fresh beaches, and *E. coli*=126/100 mL at fresh beaches |
| 2 | CA DPH, 2001 | Reclaimed water: total coliform=2.2 MPN/100 mL for disinfected tertiary recycled water and disinfected secondary 2.2 recycled water; total coliform=23 MPN/100 mL disinfected secondary 23 recycled water |
| 3 | U.S. EPA, 2002 | Potable water: detection limits for total coliform bacteria, fecal coliform bacteria, *E. coli* |

## 2.4 Frequency of Occurrence of Microorganisms in Particles

An important variable in developing particle-based standards is the frequency at which the target microorganisms are present in wastewater particles. In one study coliform bacteria were present in 0.44 to 24% of wastewater particles following biological treatment (Loge et al., 2002). The association frequency was a function of the biological treatment process used, and was related to the mean cell residence time (MCRT) within activated sludge treatment systems. Facilities with longer MCRTs had a lower percentage of particles with associated coliform bacteria. Hejkal et al. (1981) measured enteroviruses in an unchlorinated effluent at a wastewater treatment plant and found that the average concentration was 7.4 PFU/L and that only 3.4% of these viral counts were associated with solids. In the study by Hejkal et al. (1981) the particle counts were not analyzed and the frequency of association cannot be ascertained.

## 2.5 Microbial Density within Particles That Contain Microorganisms

In addition to the frequency of occurrence of microorganisms in particles, the density of the occluded microorganisms in the subset of particles containing microorganisms is also an important variable. One method for determining the density of microorganisms in particles is to observe the increase in microorganisms following chemical addition and/or physical agitation to disrupt particles and extract the occluded microorganisms (Caron et al., 2007; Ormeci and Linden, 2005; Parker and Darby, 1995). Optimal particle disruption techniques are dependent on the target microorganism, particle properties, and water quality. Verification of the effectiveness of particle disruption techniques is accomplished by observing shifts in the PSD following particle disruption.

## 2.6 Acceptable Particle Concentrations Based On Indicator Organisms

Acceptable particle concentrations were developed based on indicator organisms (total coliforms, fecal coliforms, *E. coli*, and enterococci) using the recommendations in Table 2-2 and Equation 2-1. The particle concentrations presented in this section refer to particles that are larger than 11 μm, the lower limit of particle size that can be quantified using the method employed in this study (Section 3.1.4). Equation 2-1 was used to calculate particle guidelines.

$$N_p = \left(\frac{d}{V_C}\right) \cdot \left(\frac{1}{F}\right) \cdot \left(\frac{1}{P}\right) \qquad (2\text{-}1)$$

In Equation 2-1, $N_p$ is the concentration of particles in the effluent (total number of particles/L), d is the number of target microorganisms that an individual is exposed to (microbial units/(person*exposure)), and $V_c$ is the volume of water consumed per person per day (L/person/exposure). One day is assumed to constitute a single exposure. The acceptable concentrations defined for indicator organisms are represented by the term $d/V_c$. The variable F is the fraction of particles that have occluded target microorganisms (number of particles with occluded target microorganisms/total number of particles). F ranges from 0 to 1. The variable P is the number of target microorganisms occluded per particle in the sub-domain of particles that contain one or more occluded target microorganisms (microbial units/number of particles with occluded target microorganisms). P is greater than or equal to 1.

Particle guidelines were calculated for the various indicator organisms and exposure scenarios using Equation 2-1. The microbial density that was defined above (P) was assumed as one, which corresponds with one organism per occluded particle. Different values of the frequency of association (F) were considered; the relationship between F and the particle guidelines is depicted in Figure 2-1, where different assumptions were also made regarding the microbial limits. Due to the reciprocal nature of F in the particle guidelines equation, the particle guidelines increase dramatically for small values of F. In a previous study F was observed to vary depending on the operational conditions, and values were found to vary from 0.0044 to 0.24 (Loge et al., 2002).

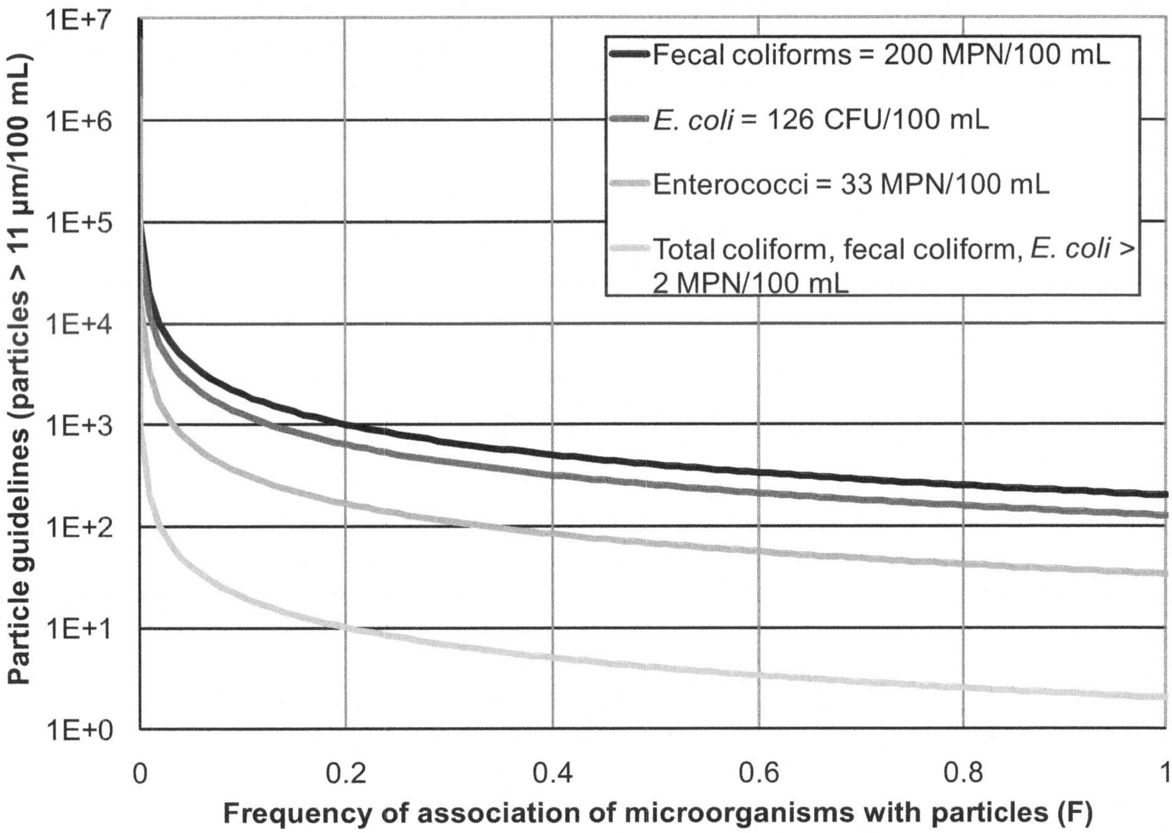


**Figure 2-1. Influence of the Frequency of Association (F) on Particle Guidelines.**

Consideration should be made for the assumptions of F and P; the values of these variables most likely vary for different microorganisms and for different treatment facilities. Data collection at the individual treatment facilities using total coliform bacteria is warranted and would result in more informed assumptions of F and P. The frequency and density of pathogens in particles probably differs from that of indicator organisms in particles, as concentrations of pathogens are typically lower than those for indicator organisms (Rose et al., 1996 and Rose et al., 2001).

In Equation 2-1 the infectious dose (d) is assumed to occur as a result of all of the microorganisms present within an occluded particle, which would require all of the microorganisms to detach from the particle and exert an individual impact. However, it is possible that the occluded microorganisms would not detach and that the occluded particle would act as a single dose of the occluded microorganism, despite the presence of multiple microorganisms.

As shown in Figure 2-1, there is a large range of values for acceptable particle concentrations. The variation depends on the target organism, exposure route, and frequency of association. Particle concentrations ranging from 2 to 45,455 particles per 100 mL were calculated when values for F of 0.0044, 0.24, and 1 were considered. A value of one was assumed for P in all cases. The most stringent standards are those calculated for a drinking water exposure route (National Drinking Water Standards) or for exposure to reclaimed water (CWRC), the lowest of which are 2 microorganisms per 100 mL. If only recreational exposure routes and accidental ingestion are considered, the lowest calculated particle guideline is 33 microorganisms per 100 mL (U.S. EPA guidelines for recreational exposure). Table 2-3 contains a summary of acceptable concentrations of particles containing target microorganisms in the plant effluent for three different exposure routes. All of the concentrations in Table 2-3 were calculated assuming that F and P are both one.

**Table 2-3. Summary of Particle Guidelines Based on Indicator Organisms.[1]**

| Microorganism | Exposure route | Particles > 11 μm/100 mL |
|---|---|---|
| *E. coli* and total coliform bacteria | Drinking water | 2 |
| Enterococci | Recreational fresh water | 33 |
| Enterococci | Recreational marine water | 35 |

[1]Particle guidelines were calculated using Equation 2-1 and are based on the assumption that F=1 and P=1.

The particle guidelines shown in Table 2-3 represent the maximum allowable particle concentrations (for particles larger than 11 μm) in the effluent to meet the effluent limits for the target microorganisms indicated, assuming that the disinfection system is adequately designed and operated to inactivate all of the dispersed target microorganisms. The particle guidelines in Table 2-3 are based on the assumption that all particles contain microorganisms. The most stringent particle guideline in Table 2-3, 2 particle/100 mL, is based on a microbial discharge limit of 2 MPN/100 mL, which is the treatment objective that was used to evaluate total coliform survival calculated using model output (Chapter 5.0).

## 2.7 Discussion and Conclusions

Particle guidelines are intended to ensure that effluents contain concentrations of occluded target microorganisms that are below the microbial limit that formed the basis of the particle guideline. Particle guidelines were calculated assuming various exposure scenarios, which impact the amount of reclaimed water incidentally ingested. Particle guidelines were also based on establishing acceptable concentrations for target microorganisms, which are indicator organisms in this chapter (pathogens are considered in Chapter 6.0). Finally, particles guidelines are also based on the frequency and density of target microorganisms in particles.

Existing regulations for indicator organisms provided the bases for particle guidelines; regulations are based on potential exposure to microorganisms in drinking water, reclaimed water, and recreational waters. Although several of the regulations presented here are based on detection limits for the analytical method used, there is some health basis for these regulations since they have historically not shown adverse health risks. However, using indicator organisms to ensure the microbiological safety of reclaimed water is problematic since the indicator organisms may not be as persistent in treatment systems when compared with pathogens. Although it is premature to calculate particle guidelines based on pathogens due to insufficient available information regarding the presence of pathogens in particles, that approach appears promising from a health risk perspective (Chapter 6.0).

Much of the uncertainty in the particles guidelines is related to uncertainty regarding the frequency of occurrence of microorganisms in particles (F) and in the microbial density within particles containing microorganisms (P). To understand F and P, information regarding microbial counts and particle concentrations must be collected concurrently for samples. There are few reports in the literature regarding F and P, and it appears that both variables have site-specific variability. In this study experiments were performed to better understand both F and P (Chapter 3.0). One way to address variability with F and P is to perform a sensitivity analyses; the information shown in Figure 2-1 was intended to demonstration the sensitivity of particle guidelines to changes in F. The variable P also varies inversely with the particle guideline concentrations, although P is always greater than one and higher P values will always result in lower particle guidelines.

A way to calculate particle guidelines based on indicator organisms that is consistent with current regulations and analytical methods is to apply the detection limits for an indicator organism (total coliform bacteria were selected for this study due to their prevalent usage in the treatment industry) and to assume values of F=1 and P=1. Assuming that all particles contain microorganisms (F=1) is conservative, but this ensures that all particles are either removed from the process flow or saturated sufficiently with chlorine to ensure inactivation. Assuming that P=1 is consistent with the assays for indicator organisms that do not differentiate between a particle containing one microorganism and a particle containing multiple microorganisms. To make the approach of P=1 more conservative, occluded microorganisms were assumed to be present in the center of particles, and intra-particle chlorine concentrations would need to be sufficiently high at all locations within the particle to inactivate the occluded microorganisms. Once the intra-particle chlorine concentrations are sufficiently high, the choice of values for P is immaterial. The particle guidelines shown in Table 2-3 (calculated using F=1) were updated based on measured values of F for the individual facilities included in this study (Section 5.3).

## CHAPTER 3.0

# BENCH-SCALE DISINFECTION TESTS AND PARTICLE CHARACTERIZATION

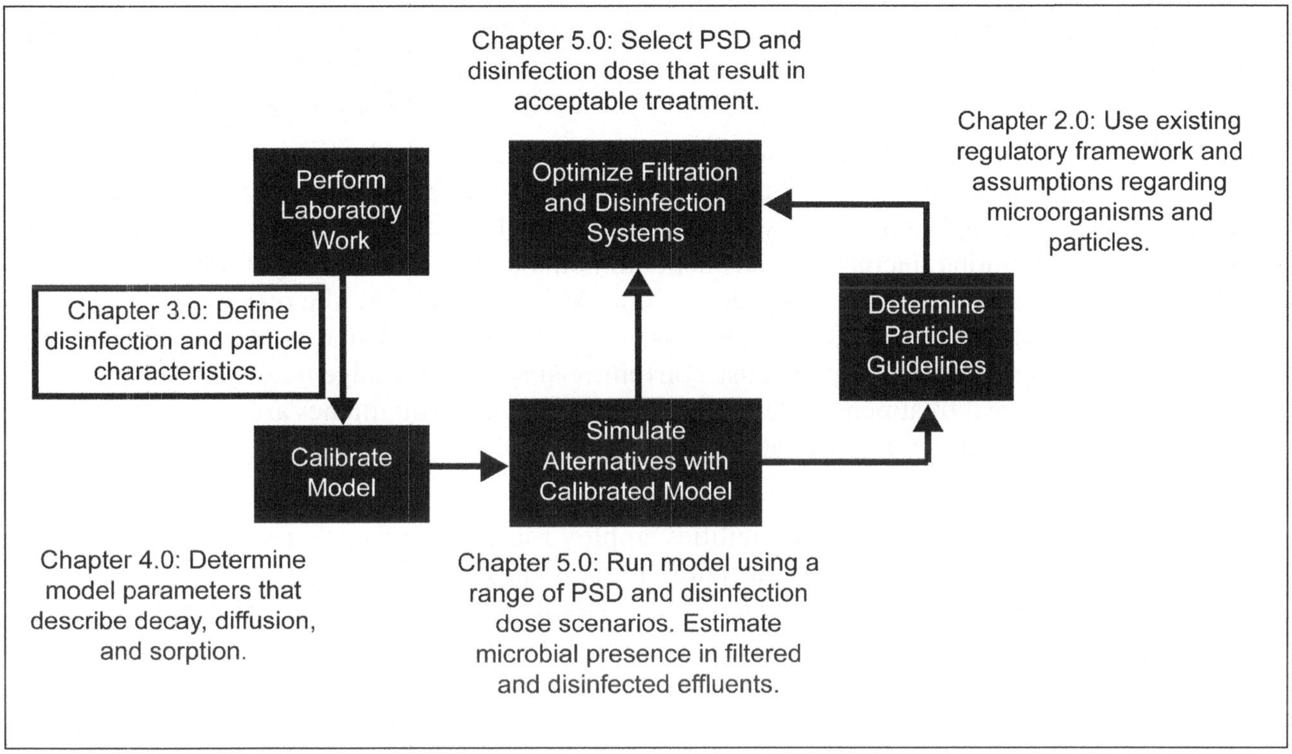


## 3.1 Introduction

Secondary effluent samples were collected from seven full-scale facilities, representing effluents derived from a variety of treatment trains. Bench-scale disinfection tests were conducted using free chlorine and combined chlorine as the disinfectants. At recorded time intervals during the disinfection tests, samples were extracted and analyzed for total or free chlorine residual and total coliform concentrations. Analysis of the PSD and wet density of particles was also performed on subsets of the samples used in the disinfection tests. The experiments were repeated at one of the facilities to better understand variability at a single facility and to validate the model and optimization approach. Particle disruption experiments were conducted for one of the facilities to better understand the density of microorganisms occluded in particles. Results of the laboratory tests, as described within this chapter, were used to calibrate the numerical model (Chapter 4.0) and run simulations (Chapter 5.0), which yielded estimates of microbial inactivation inside of particles.

## 3.2 Methods

### 3.2.1 Sample Collection

Wastewater samples were collected from seven treatment facilities. Samples were collected in sterilized containers, placed on ice, and transported to the UC Davis Civil & Environmental Engineering laboratory for immediate analysis. At each facility a secondary effluent sample was collected prior to the addition of any disinfectants and used to perform disinfection tests and particle characterization. Where available, filter effluent samples were also collected for analysis of the PSD to evaluate the impact of filtration. In some cases the filter effluent samples were not available on the day that the secondary effluent sample was collected due to intermittent operation of the filters. In some cases, separate secondary effluent and filter effluent samples were collected on a different day to evaluate the effectiveness of filtration only.

A summary of process control and operational data from the facilities is shown in Table 3-1. At four of the facilities a nitrified effluent is produced, while effluent at the other three facilities contains a substantial quantity of ammonia. Partial nitrification occurs at Facilities B, C, and D as plant conditions fluctuate and effluent ammonia concentrations vary diurnally. There is a wide range of values for mean cell residence time (MCRT) and hydraulic residence time (HRT) for the facilities. The MCRT is as low as 1.7 days at Facility F and as high as 40 days at Facility A. Although most of the plants use conventional activated sludge treatment, there is variability in biological treatment system processes used. Oxidation ditches are used at Facility A and Facility E uses a pure-oxygen aeration system.

Where filtration is used, samples were taken before and after the filtration system to access the impact on the PSD. Not all facilities employ filtration, and of those facilities that do use filtration, it is not always used for the entire process flow. Facilities E and G filter a small portion of the total flow for seasonal water reuse systems. At Facilities A, B, and F filtration is used to treat the entire waste stream. Facilities C and D do not have filtration systems. The types of filtration systems used by the facilities are variable as is chemical usage. Facility G uses continuous backwash filters, Facility A uses cloth disc filters, and Facilities E and F use a pulsed bed filter.

Most of the facilities use chlorine disinfection, relying on either free or combined chlorine, depending on the concentration of ammonia in the effluent. The disinfection process at Facility B is unique in that both free and combined chlorine are used as sequential disinfectants. At Facilities A and D, UV disinfection is used in lieu of chlorine. Data regarding the effluent ammonia concentration and the typical chlorine residual concentration and contact time are presented in Table 3-1; typical ammonia data are useful in determining whether free or combined chlorine is likely to be more important in disinfection systems. The residual chlorine concentration is variable among the plants using chlorine disinfection, ranging from less than 1 mg/L for Facility G to an average of 11 mg/L for Facility F. The chlorine disinfection data shown in Table 3-1 for Facility G pertains to the main disinfection system used to treat the majority of the waste stream. A separate disinfection system with higher residual concentrations and contact times is used to treat the side-stream that is used for the seasonal water reuse system.

Table 3-1. Process Control and Operational Data for the Facilities.

| | Facility A | Facility B | Facility C | Facility D | Facility E | Facility F | Facility G |
|---|---|---|---|---|---|---|---|
| Design flow capacity (MGD) | 10.4 | 37.5 | 1220 | 15 | 181 | 12 | 115 |
| Average annual flow rate (MGD) | 6.9 | 28.5 | 732 | 15 | 165 | 10.4 | 80 |
| Biological treatment process | Oxidation ditch | Activated sludge | Activated Sludge | Activated Sludge | Pure oxygen activated sludge | Activated sludge | Activated sludge |
| Nitrification provided? | Yes | Yes | Yes | No | No | No | No |
| MCRT (days) | 40 | 19.5 | 6-8 | 5 | 2 | 1.7 | 2.75-4.25 |
| HRT (hours) | 18 | 4.36 | 6 | 12 | 8 | NA | 3-5 |
| Aeration type | Rotors | Fine bubble diffusers | Fine bubble diffusers | Medium pore socks | Pure oxygen | Panel membranes | Fine bubble diffusers |
| Filtration | Cloth media disc | Gravity with 6 feet anthracite | None | None | Pulsed bed sand with 10-12 inches sand | Pulsed bed sand with 10-12 inches sand | Continuous backwash with 11.9 feet sand |
| Filter chemical usage | Polymer | Cationic polymer | NA | NA | Alum and polymer | NA | Alum and polymer |
| Filtration rate (gpm/sf) | NA | 3.11 | NA | NA | 3 | 1-4 | 2-3.25 |
| Disinfectant | UV | Chlorine gas followed by combined chlorine | None | UV | Combined chlorine | Combined chlorine | Sodium hypochlorite |
| Effluent ammonia concentration (mg/L) | NA | <0.5 | <0.62 | <0.5 | 26 | 5-35 | 20-35 |
| Total chlorine residual concentration (mg/L) | NA | 3.5 | NA | NA | 18 | 6.8-17 (avg.=11) | 0.3-1.0 |
| Contact time (min) | NA | 139 | NA | NA | 60 | 96-890 | 20-40 |
| CT (mg/L*min) | NA | 486 | NA | NA | | 492-5790 (avg.=1495) | NA |

### 3.2.2 Bench-Scale Chlorine Disinfection Tests

A stock solution of chlorine was prepared using purified water (Millipore, Billerica, MA) and a 4-6% (w/v) laboratory-grade solution of sodium hypochlorite (NaOCl). The pH of the stock solution was adjusted to 7.0 using hydrochloric acid (HCl). All chemicals were obtained from Fisher Scientific (Pittburgh, PA). The stock solution was standardized prior to each use by the amperometric titration method (Autocat 9000, HACH, Loveland, CO) per Standard Methods 4500-Cl.D (APHA, 2005). Measured volumes of the stock solution were added to a 1-L beaker of wastewater to produce the desired initial concentration of chlorine. Samples were withdrawn from the reactor at specified intervals and analyzed for chlorine and bacterial concentrations. Samples withdrawn for the bacteria tests were immediately dechlorinated using sodium thiosulfate per Standards Methods 9060.A (APHA, 2005). Total and free chlorine concentrations were measured using the DPD colorimetric method per Standard Methods 4500-Cl.G (APHA, 2005). Glassware used in the disinfection tests and used for measurement of chlorine (e.g. to perform dilutions) were pretreated to reduce chlorine demand by soaking in a solution of 0.005% (w/v) solution of NaOCl and rinsing five times with deionized water. Glassware used to prepare samples (e.g. filtration apparatus) and used in the bacterial assays were autoclaved to ensure sterility.

Disinfection tests on samples receiving no prior filtration were run for 90-minutes. The disinfectant residual concentration (either free or total chlorine residual) and the bacterial concentration were measured for at least eight recorded time intervals over the test period. An initial chlorine concentration was selected to provide a residual concentration of approximately 1 to 3 mg/L at the end of the test. Following the completion of the disinfectant test, measurement of the disinfectant residual continued for an additional 6 to 10 hours to provide additional data for use in model calibration.

To determine the inactivation rate coefficient for dispersed organisms, a subset of the sample was filtered serially under low vacuum through 8 μm, 5 μm, and 2 μm filters (Millipore, Billerica, MA) to remove larger solids and aggregated microorganisms. The filtered sample was used in a bench-scale disinfection test where the residual disinfectant concentration was recorded at predetermined time intervals, concurrent with sample collection for the total coliform tests. Disinfection tests with filtered samples were terminated following 10 to 20 minutes of contact time. The initial chlorine concentration was selected to yield a final chlorine residual of at least 0.5 mg/L.

All disinfection tests were conducted at room temperature and the pH was monitored to ensure that it did not change by more than 0.2 units during the test. Adjusting the pH with acid and base solutions was not necessary since the pH of the stock chlorine solution was previously altered. Controls for the disinfection tests consisted of reactors that were filled with ultra-pure water that were dosed with disinfectant (to determine the extent of volatilization) and reactors that were filled with wastewater and not disinfectant (to determine the change in microbial concentration in the absence of disinfectants).

### 3.2.3 Microbial Tests

Total coliform bacteria were used as the indicator organism in the disinfection tests. Total coliforms, fecal coliforms, enterococci, heterotrophic bacteria, indigenous F-specific coliphages, and aerobic endospores were measured in secondary effluent samples for possible use in the particle disruption tests. Total and fecal coliform bacteria were enumerated using the multiple-tube fermentation test per Standard Methods 9221 (APHA, 2005) although only the presumptive and confirmed phases were completed; the completed phase was not performed. Diluted samples were used to inoculate lauryl tryptose broth that was inoculated at 35°C. Portions of positive samples were transferred to brilliant green lactose bile broth that was incubated at 35°C to confirm the presence of total coliform. For the fecal coliform test, positive lauryl tryptose tubes were used to inoculate tubes of EC medium that were incubated at 44.5°C for 24 hours. At least three dilutions were inoculated for each sample and at least five replicates were included per dilution. The tables in APHA (2005) were used to combine all replicates and dilutions for each sample and generate a single MPN value.

Entercocci were enumerated using a multiple-tube technique per Standard Methods 9230.B (APHA, 2005). In the enterococci test, azide dextrous broth was incubated at 35°C for 48 hours with an intermediate inspection after 24 hours. Positive test tubes were used to streak Pfizer selective enterococcus agar plates that were incubated at 35°C for 24 hours. Plates were considered positive if brownish-black colonies with brown halos were present. Positive plates indicate the presence of fecal streptococci. Correctly identified colonies from the plates were transferred to brain-heart infusion broth that was incubated at 45°C for 24 hours to confirm the presence of enterococci.

Two tests were used to enumerate heterotrophic bacteria. The first test was a spread plate test using R2A agar that was incubated at 28°C for seven days. The second test used for enumeration of heterotrophic bacteria is a modification of the standard multiple-tube fermentation assay of Method 9221 B (APHA, 2005), which was used by Dietrich et al. (2007). In this technique, non-selective nutrient broth was used in place of lauryl tryptose broth and tubes were incubated at 35°C for four days. Tubes were scored as positive for growth of heterotrophic bacteria based on an increase in absorbance over sterile media measured with a spectrophotometer at a wavelength of 655 nm. The tables used to determine the MPN for the coliform tests (APHA, 2005) were also used for the multiple-tube heterotrophic test.

Male-specific coliphages were enumerated using the double agar layer method per Standard Methods 9224.C (APHA, 2005) with *E. coli* Famp as the host bacterium (ATCC 700891). Plates, consisting of typtone top and bottom agar, were incubated at 36.5°C for 16 hours. Ampicillin and streptomycin were added to the agar and to the typtone broth used to grow the host.

Aerobic endospores were enumerated per Standard Methods 9218.B, which is a membrane filtration technique (APHA, 2005). Samples temperature was first raised to 75-80°C for 10-15 minutes prior to dilution (where necessary) and membrane filtration. A minimum of 10 mL was filtered, and higher volumes were used where concentrations were lower. Following filtration, sterile membranes (Millipore, Billerica, MA) were placed on nutrient agar plates that had trypan blue added and plates were incubated at 35°C for 24 hours.

### 3.2.4 Particle Size Distribution

Two methods of PSD evaluation were used. The first method of PSD measurement involved analyzing samples using a Liquid Sampler LS-200 with a LiQuilaz-SO5-HF particle sensor (Particle Measuring Systems, Inc., Boulder, CO). The Liquilaz instrument counts the number of particles contained in 15 pre-defined particle size classes.

The second method of PSD measurement involved staining and serially filtering particles prior to collection of digital images obtained using a microscope equipped with a camera and image analysis software (Emerick et al., 1999). Particles were stained by adding 0.3% (w/v) methylene blue (Fisher Scientific, Pittburgh, PA) to samples that was first filtered (0.2 μm syringe filter) before 0.5 mL was added to a 200 mL sample. After 30 minutes of holding time at 4°C, the stained samples were sequentially filtered through 80 and 11 μm net nylon filters (Millipore, Billerica, MA). Filters were mounted on microscope slides using filtered (0.2 μm) immersion oil and a cover slip, and then were evaluated using a Zeiss Axioskop 2 plus microscope coupled with a Zeiss Axiocam camera and Zeiss Axiovision image analysis software, release 4.3 (Zeiss, Oberkochen, Germany). The radius of individual filtered and stained particles were recorded using the software, and the concentration of particles was determined by dividing the total number of particles by the volume of sample filtered. Using this microscope imaging technique, only particles larger than 11 μm can be counted since net nylon filters with pore sizes less than 11 μm are not available.

### 3.2.5 Particle Wet Density

The wet density of particles was estimated using a Percoll gradient with calibrated beads of known density (Sigma-Aldrich, St. Louis, MO). Particle wet density data were used to calculate the particle porosity (Equation 3-1), assuming that the density of solid particle material was 1.4 g/cm$^3$ (Li and Ganczarczyk, 1987).

$$\theta = \frac{\rho_s - \rho_p}{\rho_s - \rho_w} \qquad (3\text{-}1)$$

In Equation 3-1, θ is the particle porosity, $\rho_s$ is the density of the solid particle material, $\rho_w$ is the density of water, and $\rho_p$ is the wet particle density. The particle density was assumed consistent for all particles, regardless of size. Once the particle porosity was ascertained, the volumetric void ratio was calculated (Equation 3-2), which is the volume of intra-particle voids for a single particle relative to the bulk reactor volume used in the disinfection test.

$$\varepsilon_j = \frac{\theta}{V}\left(\frac{4}{3}\pi R_j^3\right) \qquad (3\text{-}2)$$

In Equation 3-2, $\varepsilon_j$ is the volumetric void ratio for the jth particle size class, $V$ is the volume of the reactor that was used in the disinfection test (1-L), and $R_i$ is the particle radius for the jth particle size class.

### 3.2.6 Frequency of Microbial Occlusion in Particles

To determine the fraction of particles that have occluded microorganisms (F), samples were first irradiated using a low dosage of UV light to inactivate the dispersed microorganisms and then the microorganisms were enumerated. A UV dose of 29 mJ/cm$^2$ was used to irradiate dispersed total coliform bacteria, as was used by Parker and Darby (1995).

A collimated beam device was used to irradiate 10 mL volume subsets of the sample in a 60 mm diameter petri dish, which was continuously stirred (NWRI, 2003; EPA 2006). Low-pressure, low-intensity UV lights (Trojan UV722, Ontario, Canada) were used and the UV intensity was measured using an IL 783A radiometer (International Light, Peabody, MA), which was calibrated by the manufacturer. UV absorbance of the samples was measured using a Cary UV-Vis spectrophotometer (Varian, Palo Alto, CA) per Standard Methods 5910.B (APHA, 2005). The UV dose was calculated using Equation C.1 in the UV Disinfection Guidance Manual (U.S. EPA, 2006, p. C-6).

### 3.2.7 Density of Occluded Microorganisms within Particles

To investigate the density of microorganisms within particles that have occluded microorganisms (P), particle disruption experiments were completed to quantify the presence of heterotrophic bacteria and aerobic endospores in particles. Similar experiments were previously completed for total coliform bacteria (Parker and Darby, 1995). The purpose of particle disruption is to break apart particles into as many smaller particles as possible and dislodge microorganisms prior to quantification to more closely quantify the total number of occluded target microorganisms.

Prior to particle disruption, samples were exposed to low dosages of UV light to inactivate the dispersed bacteria, so that only particle-associated microorganisms were enumerated. Exposure times of five minutes were used for samples analyzed for heterotrophic bacteria, which was equivalent to a UV dose of approximately 35 $mJ/cm^2$. The exposure time used for samples assayed for aerobic endospores was 10 minutes, which was equivalent to a dose of approximately 70 $mJ/cm^2$. A UV dose of 29 $mJ/cm^2$ was previously used to irradiate dispersed total coliform bacteria (Parker and Darby, 1995). In UV disinfection tests performed by Emerick et al. (1999) UV doses of approximately 25 to 50 $mJ/cm^2$ were needed to inactivate the more susceptible fraction of total coliform bacteria, and produce residual concentrations of total coliform that were in the tailing region of the inactivation curve. Results similar to those observed by Emerick et al. (1999) were confirmed in the researchers' laboratory (data not shown). The UV dose used to inactivate dispersed aerobic endospores was selected based on the inactivation information for *Bacillus subtilis* provided by Chang et al. (1985) although data for UV disinfection of aerobic spores indicates that indigenous spores in river waters can persist in samples that have undergone UV doses over 100 $mJ/cm^2$ (Caron et al., 2007; Mamane-Gravetz and Linden, 2004).

Samples were irradiated using two 10 mL petri dishes simultaneously; irradiation of samples was conducted in a manner similar to that used to measure the frequency of microbial association in particles (Section 3.2.6). Immediately following irradiation, the irradiated samples were added to a sterile container that was stored in the dark at 4°C. Once all of the samples were irradiated, the combined sample was analyzed for microbial content and subjected to the particle disruption procedures. Approximately one to three hours lapsed from the start of irradiating the samples to analysis; the delay was due to the five and 10 minute irradiation times used. Delays in analysis were minimized to prevent dispersion of occluded cells into the water column.

Two particle disruption techniques were used. In the first particle disruption technique, the following chemicals were added prior to blending the samples for 1.5 minutes at 19,000 rpm at 4°C: Tris buffer (0.01 M, pH 7.0), Zwittergent 3-12 (10-6 M), EGTA (10-3 M), and peptone (0.01%) (Parker and Darby, 1995). The second disruption technique consisted of adding Zwittergent 3-12 (100 mg/L) prior to blending at 8000 rpm for 4 min (Caron et al., 2007). A steel laboratory blender was used (Waring, No. 34BL97), which has seven settings (3,500, 7,000, 11,500, 14,500, 17,000, 19,000, and 22,000 rpm). A minimum volume of 200 mL and a maximum volume of 400 mL was blended in all experiments. A blending speed of 7,000 rpm was used with the particle disruption technique of Caron et al. (2007).

### 3.2.8 Nucleic Acid Staining

The nucleic acid stain SYTO-9 (Molecular Probes, Eugene, OR) was used to verify that filtered samples consisted of dispersed organisms. To investigate the effects of chlorination on filtered samples, the LIVE/DEAD nucleic acid stain (Molecular Probes, Eugene, OR) was used. This stain functions by staining organisms with permeable cell membranes differently than organisms with impermeable cell membranes. Presumably, a permeable cell membrane is indicative of cell damage significant enough to result in cell lysis and an intact impermeable membrane provides evidence of cell viability.

### 3.2.9 Additional Test Procedures

Turbidity was measured using a Hach 2100AN turbidimeter (Loveland, CO) according to Standard Methods 2130.B (APHA, 2005). Chemical oxygen demand (COD) was measured using a Hach DR/2500 spectrophotometer (Loveland, CO) with prior digestion. Ammonia was also measured using a Hach DR/2500 spectrophotometer (Loveland, CO). Samples were filtered with 0.2 μm sterile filters (Millipore, Billerica, MA) prior to the COD and ammonia tests.

## 3.3 Disinfection and Particle Data

Data were collected from the facilities to calibrate the model. Data in addition to that collected for the disinfection tests and the PSD are shown in Table 3-2. Samples exhibited a wide range of water quality conditions. Turbidity ranged from a low of 1.0 NTU for the sample collected from Facility A to a high of 13.3 for the sample obtained from Facility G. UV absorbance values also reflect a range of transparency characteristics; values were as low as 0.0059 for Facility A and as high as 0.4584 for Facility G. The pH range of the samples was narrow, and all samples had a pH between 7 and 8. Observed COD values were between 3 and 37.5 mg/L. The ammonia nitrogen values for the non-nitrifying facilities (Facilities E, F, and G) were high, as expected, and ranged from 28 to 39 mg/L. The ammonia nitrogen value for Facility A was not recorded, although this facility has a reliably low ammonia concentration as a result of the high MCRT (40 days). Facilities C and D were considered partially nitrifying facilities, as there were measurable amounts of ammonia in the secondary effluent samples (0.86 and 0.98 mg/L, respectively) that impacted the formation of chlorine products following the introduction of chlorine in the disinfection tests.

**Table 3-2. Water Quality and Particle Data for Secondary Effluent Samples.**

| | Facility A | Facility B | Facility C | Facility D | Facility E | Facility F | Facility G |
|---|---|---|---|---|---|---|---|
| Turbidity (NTU) | 1.0 | 1.3 | 4.3 | 8.7 | 7.0 | 4.3 | 13.3 |
| UV absorbance at 254 nm | 0.0059 | 0.11 | 0.17 | 0.20 | 0.10 | 0.24 | 0.46 |
| pH | 8.0 | 7.2 | 7.4 | 7.4 | 7.4 | 7.9 | 7.8 |
| COD (mg/L) | 3 | 19 | 30 | 18 | 30 | 35 | 38 |
| Ammonia nitrogen (mg/L) | -- | 0.02 | 0.86 | 0.98 | 28 | 29 | 39 |
| Total coliform bacteria (MPN/100 mL) | $1x10^6$ | $9x10^3$ | $5x10^4$ | $4x10^6$ | $1x10^6$ | $1x10^3$ | $2x10^6$ |
| Particles > 11 μm that contain total coliform bacteria determined by measuring the total coliform in irradiated samples (%) | 0. 41 | 0.31 | 0.015 | 0.87 | 0.21 | 0.0057 | 1.4 |
| Initial concentration of occluded total coliform as determined from the disinfection test data ($N_0$) (MPN/100 mL) | 20 | 34 | 17 | 77 | 70 | 22 | 152,200 |
| Wet density of particles (g/mL) | 1.057 | 1.048 | 1.035 | 1.043 | 1.066 | 1.056 | 1.042 |
| Particle porosity (%) | 86 | 88 | 90 | 89 | 84 | 86 | 91 |

Concentrations of total coliform bacteria differed in the samples by three orders of magnitude, ranging from $10^3$ to $10^6$ MPN/100 mL. Estimates of total coliform frequency in particles larger than 11 μm were less than 1% in all cases, and most of the values were lower than the range of 0.44 to 24% reported by Loge et al. (2002). The particle wet density data resulted in values of particle porosity that were consistent with those reported by Li and Ganczarczyk (1987), who observed values from 40 to nearly 100% with most data higher than 85%. The particle porosity values did not vary drastically, and ranged from 84% for Facility E to 91% for Facility G. The ratio of intra-particle void volume to bulk aqueous volume, also referred to as the volumetric void ratio, is dependent on the particle size of interest and is reported in Chapter 5.0 for the particle sizes used in the modeling work.

Disinfection tests were performed for all secondary effluent samples using either free or total chlorine as the disinfectant and total coliform bacteria as the indicator organism. Disinfection test conditions for the experiments are indicated in Table 3-3. For the nitrified effluents collected from Facilities A and B free chlorine was measured, while total chlorine was measured for the remainder of the facilities. Although Facilities C and D produced an essentially nitrified effluent, there were measurable amounts of ammonia (0.86 and 0.98 mg/L, respectively) that resulted in considerable production of combined chlorine following the addition of chlorine, as was confirmed by a comparison of free and total chlorine residuals (data not shown).

**Table 3-3. Disinfection Test Conditions and Results for Tests with Filtered and Unfiltered Secondary Effluent Samples.**

| | Facility A | Facility B | Facility C | Facility D | Facility E | Facility F | Facility G |
|---|---|---|---|---|---|---|---|
| Type of chlorine residual measured | Free | Free | Total | Total | Total | Total | Total |
| Chlorine added, disinfection test performed without prior filtration (mg/L) | 8 | 6 | 6 | 4 | 20 | 12 | 1.5 |
| Instantaneous chlorine demand in the test performed on an unfiltered sample (mg/L) | 3.1 | 3.1 | 2.0 | 0.6 | 0.0 | 0.8 | 0.5 |
| Chlorine added, filtered (2 μm) water disinfection test (mg/L) | 2 | 1 | 2 | 3 | 5 | 7 | 1.5 |
| Total coliform inactivation rate coefficient for dispersed microbes passing through a 2 μm filter[1] ($k_i$) (L/mg*min) | 14 | 8.2 | 0.17 | 0.55 | 0.11 | 0.23 | 0.094 |

[1]Inactivation coefficients were calculated using data from the disinfection tests performed using filtered (2 μm) secondary effluent samples.

The initial concentration of chlorine added to the reactors was variable; selection of an appropriate concentration depended on factors such as water quality and inherent resistance of the total coliform bacteria at each facility. The initial chlorine concentrations were selected based on input from representatives at each facilities as well as demonstration of the chlorine concentration needed to maintain a residual throughout the entire test. Typically more than one disinfection test was performed on any given day; however, many of the data were not useful for model calibration because the inactivation was either too high or too low. An appropriately designed disinfection test, in the context of providing data for model calibration, is one that produces an inactivation curve with a substantial tailing region. Selection of an appropriate initial chlorine concentration often required several attempts. In the disinfection tests the initial chlorine concentration ranged from 1.5 mg/L for Facility G to 20 mg/L for Facility E. The concentration of chlorine required to inactive dispersed microorganisms in the filtered water disinfection tests was much lower than in the disinfection tests performed on samples without prior filtration, and ranged from 1 mg/L for Facility B to 7 mg/L for Facility F. The instantaneous chlorine demand was determined manually by viewing the chlorine residual concentrations that were observed during the disinfection tests. The instantaneous chlorine demand was the difference between the chlorine added to the reactors and the chlorine that was measured in the first few minutes of the test; the instantaneous chlorine concentration was adjusted during model calibration to obtain the best-fit (described in Chapter 5.0).

Results from the disinfection tests using samples without prior filtration are shown in Figure 3-1. The initial concentration of total coliform bacteria rapidly decreased during the first ten minutes of the disinfection tests. In most cases, the rate of inactivation during the initial part of the test (less than 10 minutes of contact time) was linear on a logarithmic scale. After approximately 10 minutes of contact time, the inactivation curve became relatively flat, signifying slow rates of inactivation. An inactivation curve with a tailing region was achieved for all samples used in the disinfection tests. In this study it was assumed that the initial region of the inactivation curve represents inactivation of dispersed microorganisms and that the tailing region of the inactivation curve represents inactivation of microorganisms contained within particles. The data for Facility G are much higher than data for the other facilities; however, the degree of inactivation for each sample is largely a function of the initial concentration of chlorine added to the reactors. A higher initial chlorine concentration for the disinfection test performed with the sample from Facility G would have produced an inactivation curve with a lower tailing concentration.

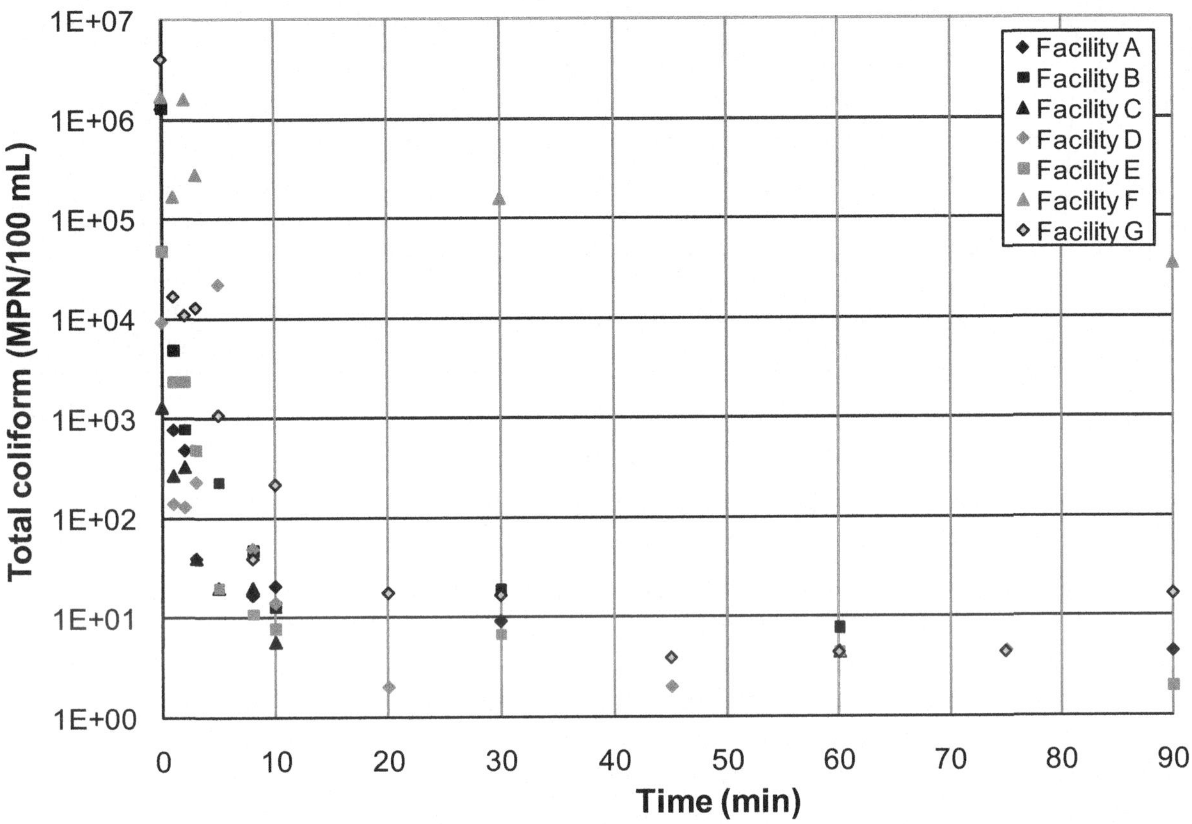


**Figure 3-1. Results for Chlorine Disinfection Tests Conducted Using Unfiltered Secondary Effluent Samples.**

In addition to the disinfection tests performed with unfiltered secondary effluent samples, disinfection tests were also performed using filtered (2 μm) secondary effluent samples as shown in Figure 3-2. The purpose of the disinfection tests using filtered secondary effluent samples was to determine the inactivation coefficient for dispersed bacteria, which is assumed to be equivalent to the inactivation coefficient for bacteria occluded in particles. Dispersion of the filtered samples was verified by staining samples prior to viewing them with an epi-fluorescent microscope, which verified the effectiveness of the filtration procedure.

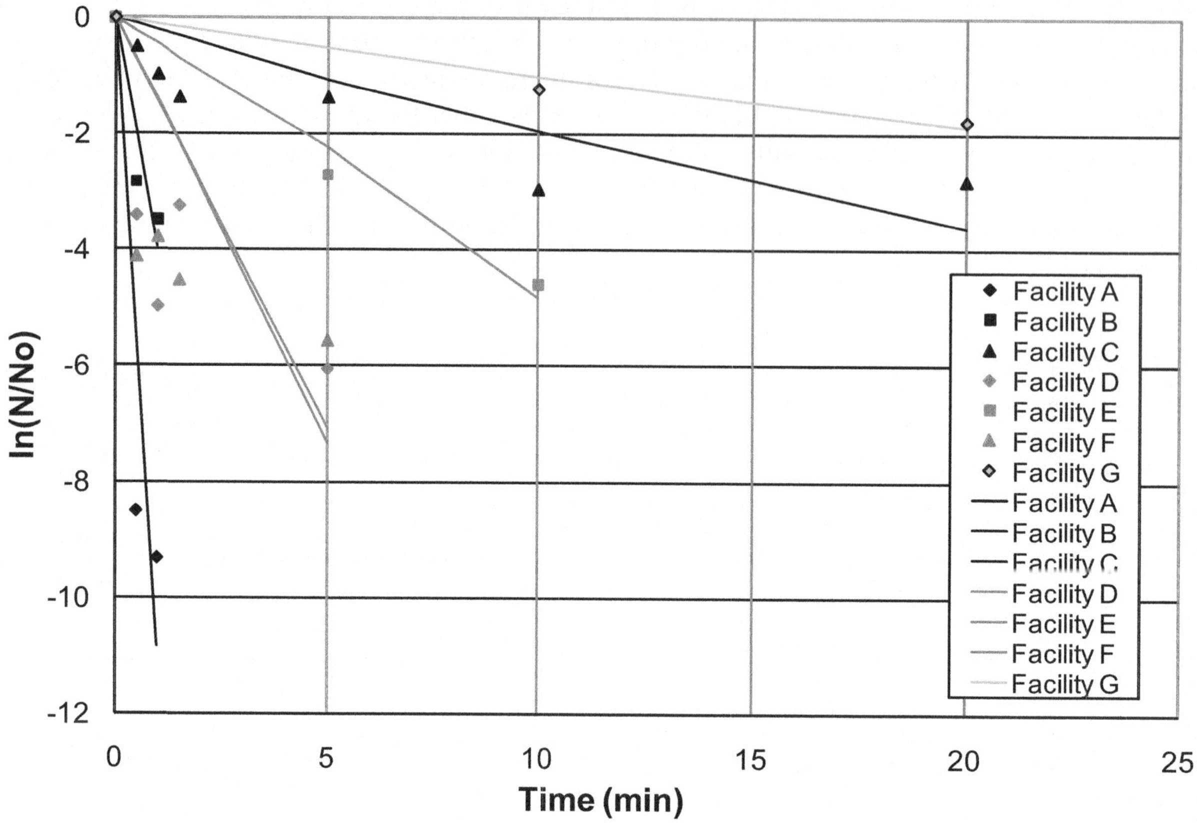


**Figure 3-2. Results for Chlorine Disinfection Tests Conducted Using Filtered (2 µm) Secondary Effluent.**

Equation 1-2 was used to model the data, assuming first-order decay kinetics, and the results are shown in Figure 3-2. Inactivation rate coefficients for all facilities are shown in Table 3-3. Where free chlorine was used as the disinfectant for the nitrified samples from Facilities A and B, the inactivation rates were much higher than for the other samples. The inactivation rates observed at Facilities A and B were 13.9 and 8.20 L/mg*min, respectively, whereas the inactivation rate coefficients for the other facilities ranged from 0.0938 to 0.548 L/mg*min.

PSD data for all facilities were divided into four particle size classes as shown in Table 3-4. The total concentrations of particles larger than 11 µm were highly variable, ranging from nearly 5000 particles per 100 mL for Facility A to 160,000 particles per 100 mL for Facility C. In all cases, the smallest particle size class contains the highest concentration of particles and the largest size class contains relatively fewer particles, although the distribution of particle sizes varied amongst the facilities. To provide more detailed information on PSD, histograms of the particle data are contained in Appendix B. Turbidity data are also shown in Table 3-4 to demonstrate that higher turbidity values do not necessarily correspond with higher particle counts. For example the turbidity of the sample from Facility B was higher than the turbidity of the sample from Facility A although the particle concentration was lower. A similar comparison between the data from Facilities C and D is even more dramatic. One conclusion of the data shown in Table 3-4 is that even secondary effluents with very low turbidities still have a lot of particles (thousands per 100 mL). The facilities with the highest particle counts, Facilities C and G (160,000 and 150,000 particles larger than 11 µm per 100 mL), have very different turbidities (4.3 and 13.3 NTU).

**Table 3-4. Particle Size Distribution Data for Secondary Effluent Samples, with Turbidity Data Shown for Comparison.**

| Range of particle sizes (μm) | Particles in each particle size class (%) | | | | | | |
|---|---|---|---|---|---|---|---|
| | Facility A | Facility B | Facility C | Facility D | Facility E | Facility F | Facility G |
| 11-40 | 60.2% | 92.3% | 94.3% | 92.6% | 95.4% | 80.6% | 86.7% |
| 40-80 | 37.0% | 2.4% | 5.3% | 6.8% | 4.4% | 16.0% | 10.6% |
| 80-120 | 2.3% | 1.8% | 0.4% | 0.4% | 0.2% | 2.6% | 1.8% |
| >120 | 0.4% | 3.5% | 0.1% | 0.1% | 0.0% | 0.8% | 0.9% |
| **Total (# of particles > 11 μm/100 mL)** | **4900** | **3000** | **160,000** | **81,000** | **82,000** | **63,000** | **150,000** |
| Turbidity (NTU) | 1.0 | 1.3 | 4.3 | 8.7 | 7.0 | 4.3 | 13.3 |

Representative images of particles collected with a microscope are contained in Appendix C. The images portray the variability in particle shapes and sizes. When counting the particles, a systematic approach was necessary. Although a computer program was used to identify particles and record information regarding the size and shape of particles, interpretation of the data was necessary to determine which particles to include in the count (e.g. algae were not counted). Interpretation of the images was also necessary to ensure that closely located particles were not counted as one and that an apparent single particle was not counted as multiple particles (e.g. as a result of transparent portions of a particle). Most particles were easily counted using the microscope and the imaging software although all counted particles were viewed prior to inclusion in the particle database for each sample.

As shown in the images in Appendix C, many of the particles were easily identifiable such as the relatively distinct particles shown in Figure C-1 and C-2 for Facility A and in Figures C-7 and C-8 for Facility C. The particles shown for Facility C were less distinct and contained regions that were more transparent (Figures C-1 and C-3). In general, particles found in the tertiary effluent samples tended to have similar characteristics as the particles found in the secondary effluent samples (see examples taken from Facility B in Figures C-5 and C-6). The image shown in Figure C-9 indicates that the sample contained algae in addition to effluent wastewater particles. The algae were predominantly captured by the 80 μm filter and were not present on the 11 μm filter (Figure C-10). Particles indentified from Facility E tended to be stringier than particles from other facilities (Figure C-11 and C-12). One of the particles shown in Figure C-13 is quite large and demonstrates that large particles can be present in effluents; this particle also appears to be aggregating with other particles.

## 3.4 Particle Disruption Experiments to Determine Microbial Density in Particles

Prior to conducting the particle disruption experiments, samples were collected from Facility A and were assayed for several indicator organisms (Figure 3-3). The data shown in Figure 3-3 represent averages for at least two samples. In all cases, results of replicate samples varied by less than an order of magnitude. The purpose of enumerating indicator organisms was to determine the best candidates for the particle disruption experiments. The indicator organisms tested were aerobic endospores, enterococci, coliphage, total coliforms, fecal coliforms, *E. coli*, and heterotrophic bacteria. All indicator organisms were reliably present in untreated wastewater and in secondary effluent with the exception of coliphage, which was not found in the secondary effluent samples. Based on the results shown in Figure 3-3, aerobic endospores and heterotrophic bacteria were selected for use in particle disruption experiments. Aerobic endospores were selected as a result of their suggested use as surrogates for protozoan pathogens. Heterotrophic bacteria were selected due to their high concentrations; it was assumed that culturable heterotrophic bacteria would also be present in wastewater particles at relatively high concentrations.

Results of the particle disruption experiments using aerobic endospores are shown in Figure 3-4. Concentrations of aerobic endospores are shown for undisinfected secondary effluent samples, secondary effluent samples that had been exposed to a low dose of UV light, and irradiated samples that had undergone particle disruption. The error bars shown in Figure 3-4 represent the 95 percentile confidence interval. Experiments were completed four times; three of the experiments (labeled 'a', 'b', and 'c' in Figure 3-4) were completed using the particle disruption technique described by Parker and Darby (1995) and one of the experiments (labeled 'd' in Figure 3-4) was completed using the disruption technique described by Caron et al. (2007). In all four experiments, the low dose of UV light reduced the quantity of aerobic endospores, but did not completely inactivate the organisms present. For the experimental results labeled 'a' and 'b' in Figure 3-4, the concentration of aerobic endospores was not significantly different following particle disruption ($p>0.05$ in a t-test). In the other two experiments, the concentration of aerobic endospores significantly decreased ($p<0.05$) following particle disruption although the decrease was less than an order of magnitude.

The results of the particle disruption experiments using hetereotrophic bacteria are shown in Figure 3-5. Similar to data shown in Figure 3-4, concentrations are shown for heterotrophic bacteria found in undisinfected secondary effluent samples, secondary effluent samples that have been exposed to a low dose of UV light, and irradiated samples that have undergone particle disruption. Again, the error bars represent the 95 percentile confidence interval. Particle disruption experiments were completed twice for heterotrophic bacteria; the experiment was completed once (labeled 'a' in Figure 3-5) using the particle disruption technique described by Parker and Darby (1995) and repeated (labeled 'b' and 'c') using the disruption technique described by Caron et al. (2007). In the second experiment, two hetertrophic bacterial assays were used to determine if there was a difference between the two assays. When the particle disruption technique of Parker and Darby (1995) was used, the concentration of hetertrophic bacteria significantly increased following particle disruption. When the disruption technique of Caron et al. (2007) was used, the concentration of heterotrophic bacteria did not significantly change following particle disruption, and the choice of heterotrophic bacteria assay did not affect this outcome.

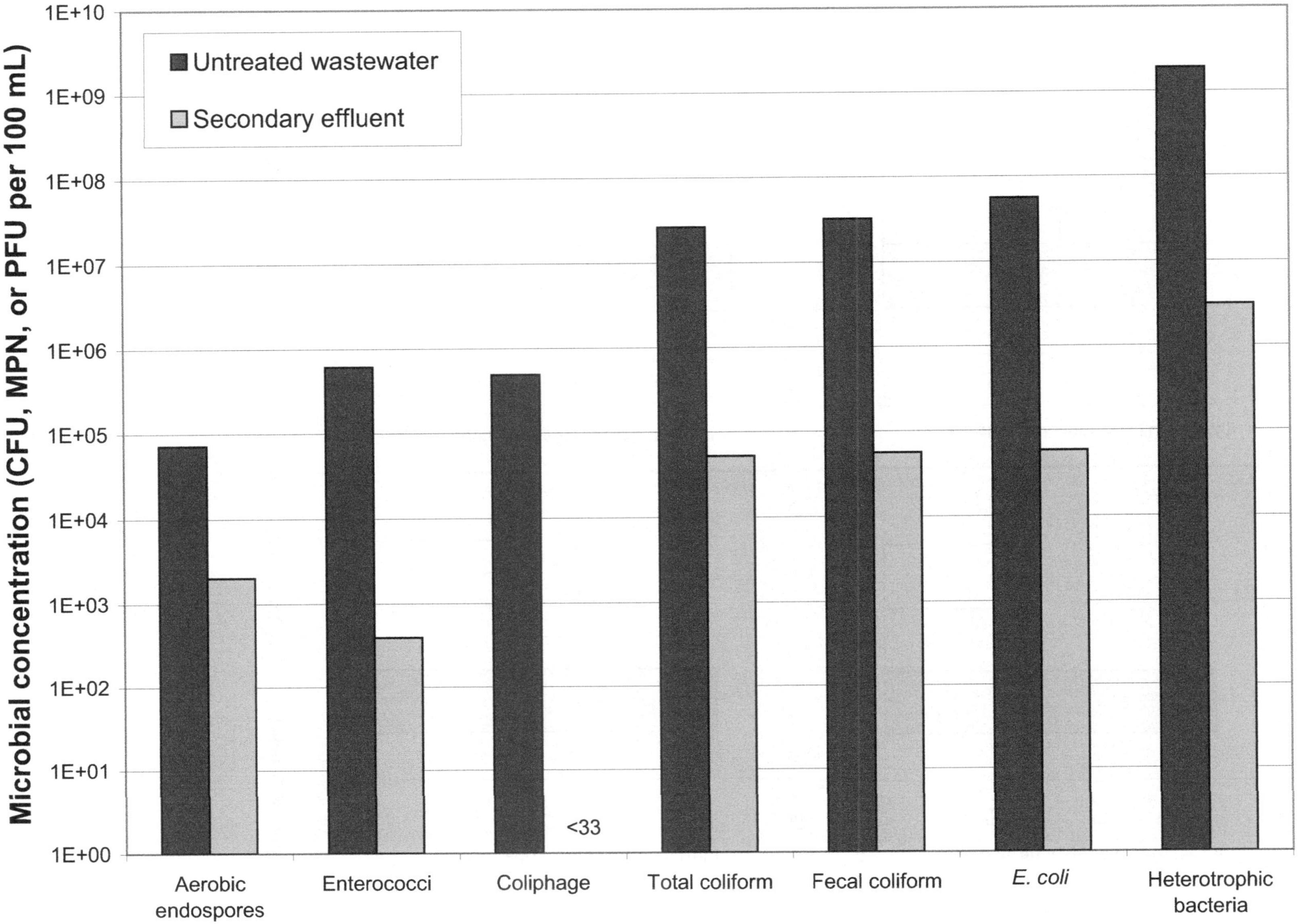


**Figure 3-3. Presence of Indicator Organisms in Facility A Untreated Wastewater and Secondary Effluent Samples.**

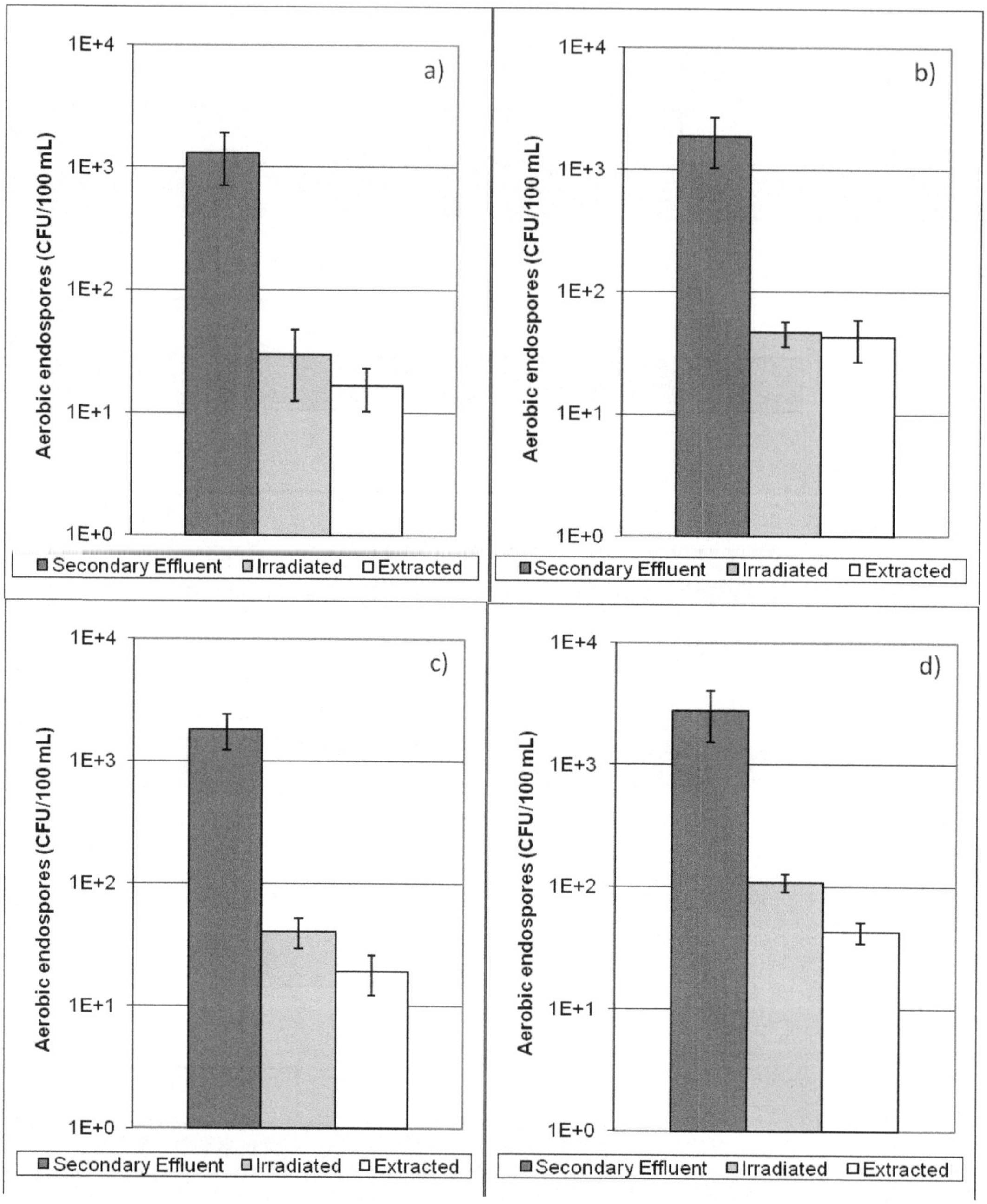


**Figure 3-4. Results of Particle Disruption Experiments Using Aerobic Endospores and Facility A Secondary Effluent.**

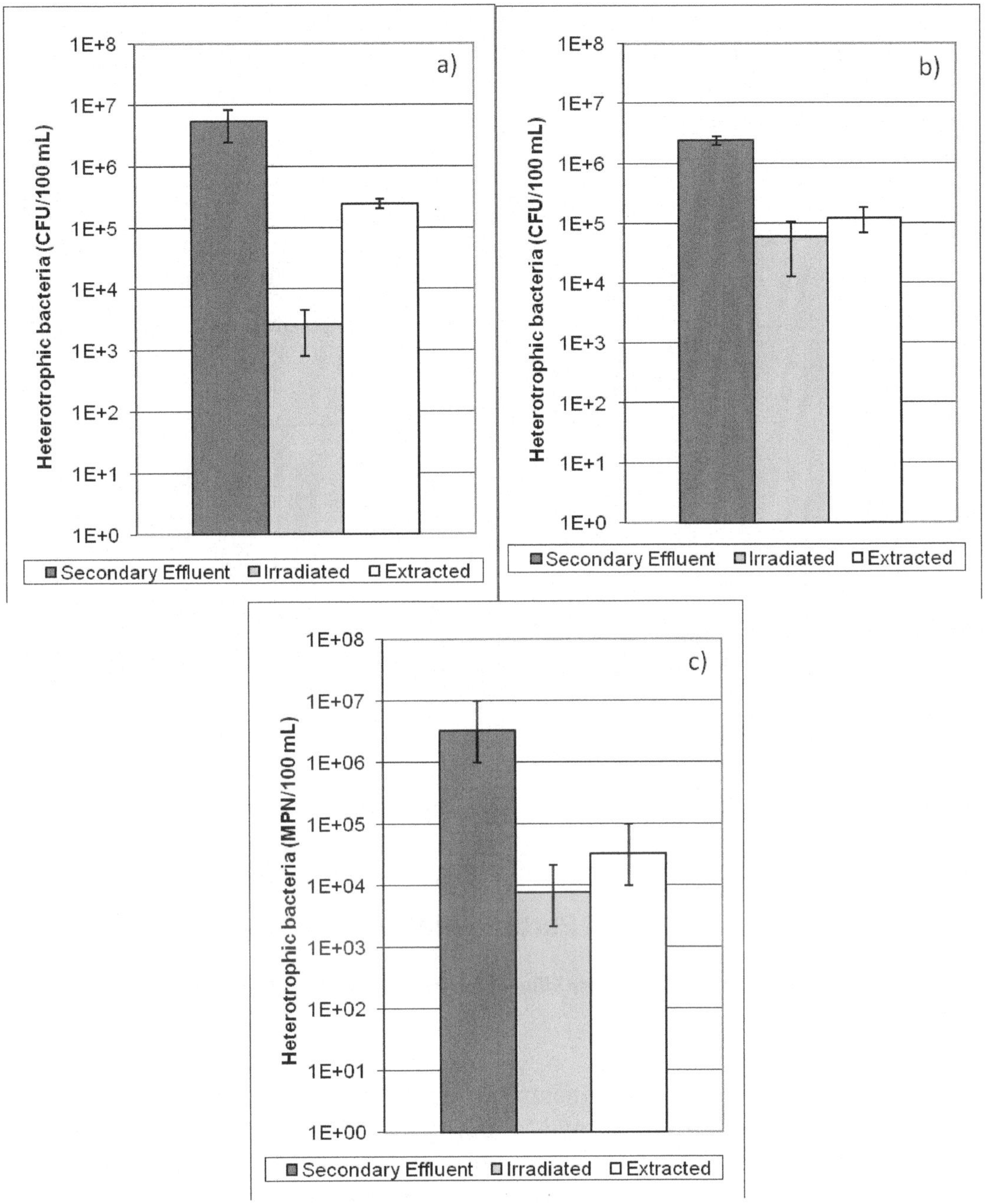


**Figure 3-5. Results of Particle Disruption Experiments Using Heterotrophic Bacteria and Facility A Secondary Effluent.**

In the first experiment using heterotrophic bacteria (labeled 'a' in Figure 3-5) and the third experiment using aerobic endospores (labeled 'c' in Figure 3-4), the effectiveness of the particle disruption technique was verified by analyzing the PSD before and after particle disruption as shown in Figure 3-6. A paired t-test revealed that the particle concentrations before and after particle disruption were significantly different ($p<0.05$) for every bin except the bins where the upper range of the particle diameter was 6.727 µm and 8 µm. For most particle size classes, the concentration of particles increased following particle disruption, except that the concentration of particles with diameters less than 0.5 µm decreased following particle disruption. For the three largest particle size classes, the concentration of particles decreased following particle disruption. The increases in the number of small particles and decreases in the number of large particles shown in Figure 3-6 were expected because particle disruption is intended to break up larger particles into smaller particles.

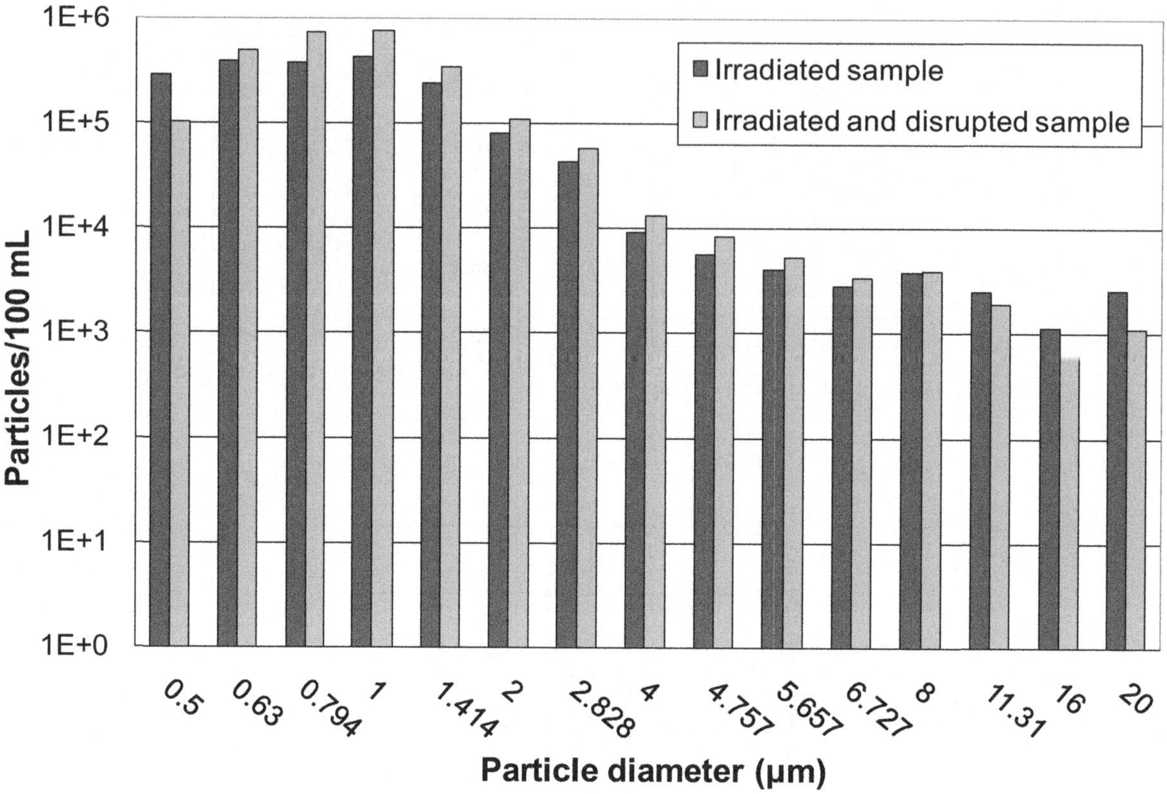


**Figure 3-6. Particle Size Distribution of a Secondary Effluent Sample from Facility A, Before and After Particle Disruption.**

For the particle disruption experiments performed using the technique of Caron et al. (2007), labeled 'd' in Figure 3-4 and 'b' and 'c' in Figure 3-5, the PSD was evaluated using the microscope counting method for the secondary effluent sample. The total number of particles larger than 11 μm was 4,100/100 mL. Previous measurements of PSD for this facility indicate typical particle counts ranging from 4000 to 7000/100 mL (data not shown). Assuming that the microorganisms surviving low doses of UV light represent the occluded fraction of microorganisms, an estimated 3% of particles contain aerobic endospores, as there were 110 CFU/100 mL aerobic endospores in the irradiated sample. The absence of an increase for aerobic endospores following particle disruption indicates that either the particle disruption technique was not effective, or the density of aerobic endospores wastewater particles is approximately one.

The concentrations of heterotrophic bacteria measured in irradiated samples using the multiple-tube and the plate count assays were different by an order of magnitude (7900 MPN/100 mL and 59,000 CFU/100 mL) for the same sample. The values are also 2 to 14 times higher than the total particle counts for particles larger than 11 μm, indicating that not all of the dispersed heterotrophic bacteria were inactivated with the selected dose of 35 mJ/cm$^2$. It is also possible that heterotrophic bacteria were occluded in particles smaller than 11 µm, the lower limit of the particle size for the PSD method used. If the frequency of association is assumed as one, meaning that each particle contains at least one heterotrophic bacterium, than 3800 MPN/100 mL and 54,900 CFU/100 mL dispersed heterotrophic bacteria were mistakenly counted as occluded heterotrophic bacteria or were occluded in particles smaller than 11 µm. Following particle disruption, the quantity of heterotrophic bacteria increased to 33,000 MPN/100 mL and 125,000 CFU/100 mL. If the concentration of dispersed heterotrophic bacteria that were counted as occluded heterotrophic bacteria were subtracted from the concentrations of heterotrophic bacteria enumerated in irradiated samples, the release ratio (P) for the particle disruption experiments is 7.1 based on the multiple-tube test and 17.1 based on the spread plate test. The release ratio is the quantity of microorganisms in the irradiated and extracted sample divided by the quantity in the irradiated sample; the release ratio is synonymous with the microbial density occluded in the subset of particles containing microorganisms (P).

When particle disruption experiments were performed for heterotrophic bacteria using the technique of Parker and Darby (1995) (labeled 'a' in Figure 3-5), the concentration of heterotrophic bacteria in the irradiated sample was 2700 CFU/100 mL. Following particle disruption, the concentration of heterotrophic bacteria increased to 250,000 CFU/100 mL, and the ratio of heterotrophic bacteria in the irradiated and extracted sample relative to that in the irradiated sample was 93. The PSD was not analyzed for this sample, however, the value of heterotrophic bacteria measured in the irradiated sample, 2700 CFU/100 mL, is lower than the concentration of particles typically observed for this facility (4000 to 7000 particles larger than 11 µm per 100 mL). For this sample it is not clear that the low UV dose was insufficient to inactivate the dispersed fraction of heterotrophic bacteria.

## 3.5 Disinfection and Particle Data for the Model Simulation Exercise

Additional data were collected from Facility A to calibrate the radial pore diffusion model. Disinfection tests and particle characterization were repeated three times. Experiment A-1 refers to the work that was done originally to calibrate the model and conduct the optimization protocols; Experiments A-2, A-3 and A-4 refer to the experimental work that was done for the model simulation exercise. Water quality and particle data that were collected for the model simulations exercise are shown in Table 3-5. The water quality data are fairly consistent although all samples were collected during the dry season from June to September and all samples were collected in the morning. The values for turbidity, UV absorbance, and pH were fairly consistent for the four sampling dates, although the UV absorbance for the first sample collected was an order of magnitude lower than the other values. The COD concentration varied from non-detectable values up to 16 mg/L. When it was tested, the ammonia nitrogen concentration was always below the detection limit. The concentration of total coliform differed by two orders of magnitude for the samples, ranging from $10^4$ to $10^6$ MPN/100 mL. The frequency of total coliform bacteria in particles ranged from 0.1 to 2.0%, values that were not entirely inconsistent with the range of values of 0.44 to 24% reported by Loge et al. (2002). To conserve resources for the disinfection tests, the particle wet density test was only completed for Experiment A-1 and a particle porosity of 0.9 was assumed in subsequent experiments, which is consistent with values reported by Li and Ganczarczyk (1987).

**Table 3-5. Model Simulation Exercise Water Quality and Particle Data.**

| | Experiment A-1 | Experiment A-2 | Experiment A-3 | Experiment A-4 |
|---|---|---|---|---|
| Turbidity (NTU) | 1.0 | 1.1 | 0.91 | 1.3 |
| UV absorbance at 254 nm | 0.0059 | 0.095 | 0.086 | 0.083 |
| pH | 8.0 | 7.9 | 7.8 | 7.9 |
| COD (mg/L) | 3 | 16 | <3 | 4.5 |
| Ammonia nitrogen (mg/L) | -- | <0.02 | <0.02 | <0.02 |
| Total coliform bacteria (MPN/100 mL) | $1x10^6$ | $3x10^5$ | $8x10^4$ | $2x10^5$ |
| Particles > 11 μm that contain total coliform bacteria determined by measuring total coliform bacteria in irradiated samples (%) | 0. 4 | 2.0 | 0.1 | 0.7 |
| Initial concentration of occluded total coliform bacteria determined by observing disinfection data and extending a best-fit line through the tailing data (MPN/100 mL) | 20 | 194 | 86 | 312 |

Multiple disinfection experiments were conducted for the model simulation exercise as shown in Table 3-6. In all cases free chlorine was measured since the ammonia concentration was so low that formation of combined chlorine products was expected to be minimal. For the disinfection tests without prior filtration, a variety of initial chlorine concentrations were used, and it was found that the chlorine demand in Experiments A-2, A-3, and A-4 was lower than in the original Experiment A-1. Additional disinfection tests were also conducted using a subset of the sample that had undergone filtration with a 20 μm net nylon filter to remove the largest particles. The total coliform inactivation rate coefficients determined using data from the disinfection tests performed with filtered samples were much higher than values observed for other facilities (Table 3-3).

**Table 3-6. Model Simulation Exercise Disinfection Test Conditions and Results.**

| | Experiment A-1 | Experiment A-2 | Experiment A-3 | Experiment A-4 |
|---|---|---|---|---|
| Type of chlorine residual measured | Free | Free | Free | Free |
| Chlorine added, disinfection test performed without prior filtration (mg/L) | 8 | 8, 6 | 5, 3, 7 | 4, 3, 5 |
| Instantaneous chlorine demand in the test performed on an unfiltered sample (mg/L) | 3.1 | 4 (8 mg/L test) | 2.2 (5 mg/L test) | 2.3 (4 mg/L test) |
| Chlorine added, filtered (20 μm) water disinfection test (mg/L) | -- | 8 | 5 | 4 |
| Chlorine added, filtered (2 μm) water disinfection test (mg/L) | 2 | 2 | 1, 1.5 | 0.8, 1 |
| Total coliform inactivation rate coefficient for microorganisms smaller than 2 μm ($k_i$) (L/mg*min) | 14 | 23 | 20 | 35 |

Results from the disinfection tests using secondary effluent samples from Facility A without prior filtration and with filtration using a 20 μm filter are shown in Figure 3-7. In most disinfection tests a tailing region was achieved in the inactivation curve, and there were sufficient data for model calibration and the model simulation exercise. There was variability in chlorine demand for samples collected on different days. In Experiments A-1 and A-3 an initial chlorine concentration of 8 mg/L was sufficient to generate a tailing region; however, in Experiment 2 the concentration of 8 mg/L was too high to generate tailing and the total coliform counts for samples collected after 30 minutes were below detection limits. For Experiments A-2 and A-3, disinfection tests with samples that had undergone filtration with 20 μm filters did not result in a tailing region as most of the total coliform assays completed for the tests yielded results below the detection limit. In Experiments A-2 and A-3, initial chlorine concentrations of 6 and 5 mg/L were used, respectively, for filtered (20 μm) samples. In Experiment A-4 an initial chlorine concentration of 4 mg/L resulted in an inactivation curve with tailing for the filtered (20 μm) sample.

The instantaneous chlorine demand values reported in Table 3-6 were determined by viewing the chlorine residual data collected in the disinfection tests. The instantaneous chlorine demand is the initial concentration of chlorine added to the reactors subtracted by the concentration of chlorine measured in the first few minutes of the disinfection test. The instantaneous chlorine demand appears to be a function of the initial concentration of chlorine added in the disinfection test, and it is higher when more chlorine is added to the reactors. Data were not available for all of the disinfection tests because data were not collected in the first 10 minutes of some of the tests because only the tailing region of the test was of interest.

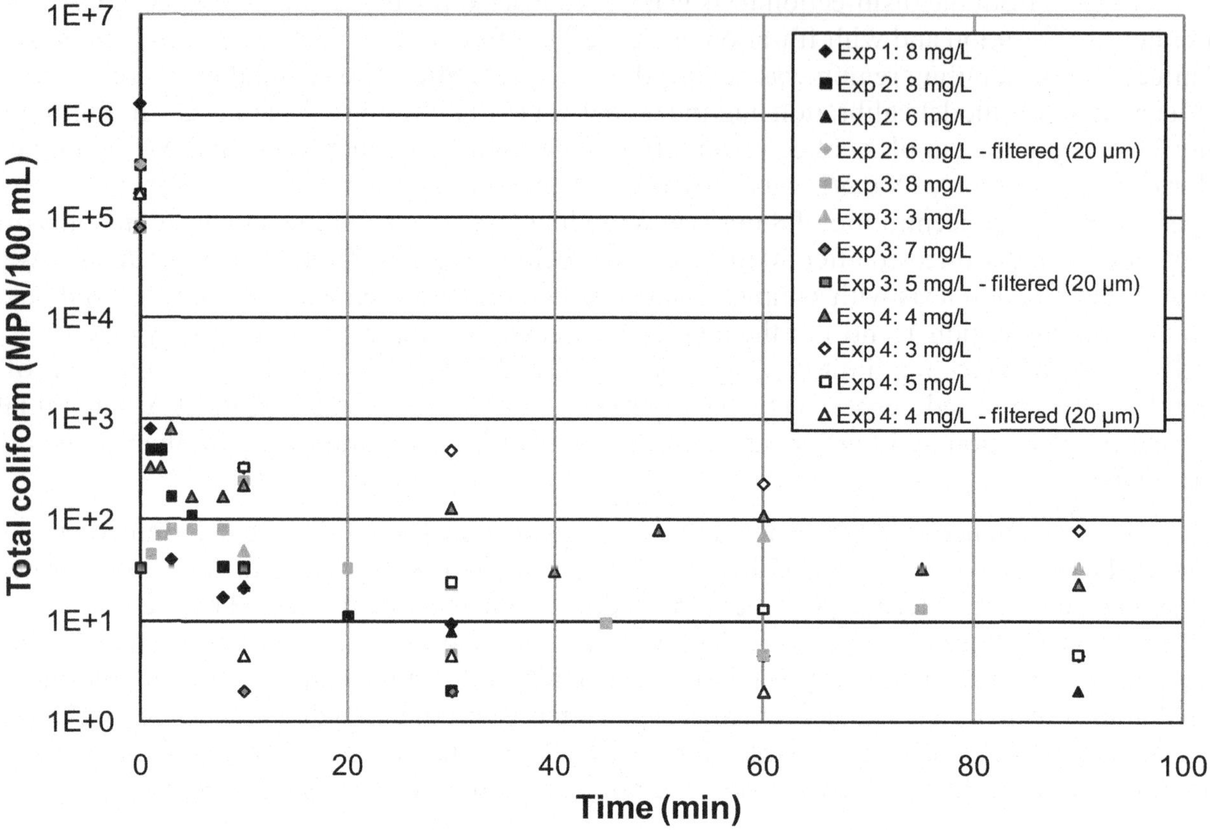


**Figure 3-7. Model Simulation Exercise Chlorine Disinfection Test Results for Unfiltered Secondary Effluent Samples from Facility A.**

Results from the disinfection tests using filtered (2 µm) secondary effluent samples collected from Facility A are shown in Figure 3-8. As shown in the figure, the data are nearly linear on a logarithmic scale, indicating that a first-order decay equation is appropriate. The lines shown in Figure 3-8 are the log-linear models that were developed based on a best-fit for the data (Equation 1-2). Inactivation of dispersed total coliform was rapid, and concentrations decreased to levels below detection limits within two minutes after the addition of chlorine. In the disinfection tests performed using filtered (2 µm) samples, it was challenging to select an initial chlorine dose that resulted in a measurable chlorine residual and in a measurable concentration of total coliform bacteria.

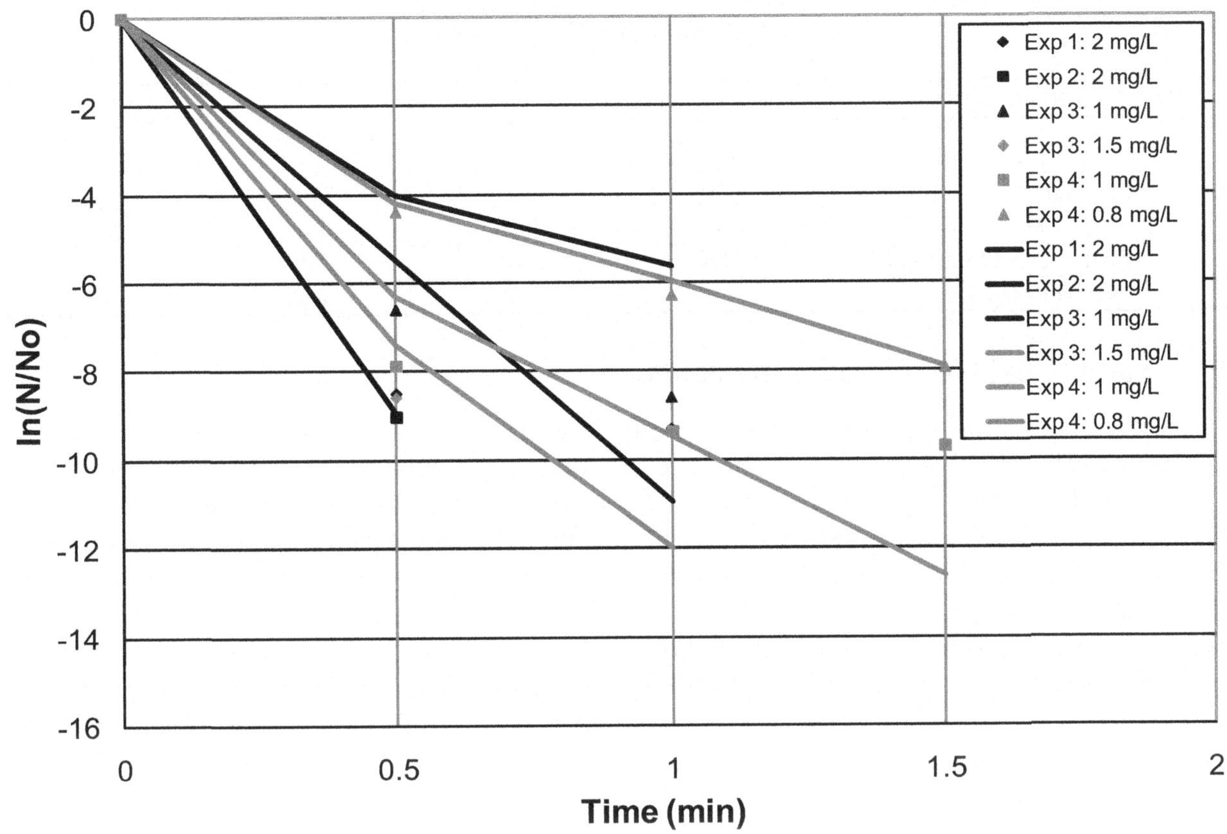


**Figure 3-8. Model Simulation Exercise Chlorine Disinfection Test Results for Filtered (2 μm) Secondary Effluent Samples from Facility A.**

Particle data collected for the model simulation exercise are shown in Table 3-7. The total particle counts for Experiments A-1 and A-2 were similar: 4900 and 4100 particles per 100 mL, respectively for particles with diameters larger than 11 μm. The total particle counts for Experiments A-3 and A-4 were the same, 7100 particles per 100 mL, but the counts were much higher than what was observed for the first two experiments. The sample for Experiment A-1 had a lower percentage of small particles relative to what was observed for the other samples. A sample collected from Facility A during the particle disruption experiments had a total particle count for particles larger than 11 μm of 4100/100 mL. When comparing the water quality data shown in Table 3-5 with the data shown in Table 3-7, it is apparent that there is no clear trend between particle counts and water quality parameters such as turbidity and COD.

Table 3-7. Model Simulation Exercise Particle Size Distribution Data for Secondary Effluent Samples, with Turbidity Data Shown for Comparison.

| Range of particle sizes (μm) | Experiment 1 | Experiment 2 | Experiment 3 | Experiment 4 |
|---|---|---|---|---|
| 11-40 | 60.3% | 91.6% | 91.2% | 97.1% |
| 40-80 | 37.0% | 7.1% | 7.6% | 2.6% |
| 80-120 | 2.3% | 1.2% | 1.1% | 0.2% |
| >120 | 0.4% | 0.1% | 0.2% | 0.1% |
| **Total (# of particles > 11 μm/100 mL)** | **4900** | **4100** | **7100** | **7100** |
| **Turbidity (NTU)** | **1.0** | **1.1** | **0.91** | **1.3** |

## 3.6 Particle Data for Filtration Systems

Where filtration systems were in place and operating, tertiary samples were collected to evaluate the effectiveness of filtration. In the case of Facility E, collection of a tertiary sample was not possible because the facility operates a seasonal filtration system that was not in operation when samples were collected from this facility. The PSD data for filter influent and effluent samples collected from Facility A are shown in Figure 3-9; filter effluent samples were not available on the day that a sample was collected for the optimization protocols, so filter influent and effluent samples were collected on a different day for PSD analysis only. Facility A uses cloth disc filters. Facility B uses a deep-bed gravity filter with anthracite media; the PSD data for this facility are shown in Figure 3-10. Facility F uses a pulsed-bed sand filter; PSD data collected for filter influent and effluent samples is shown in Figure 3-11.

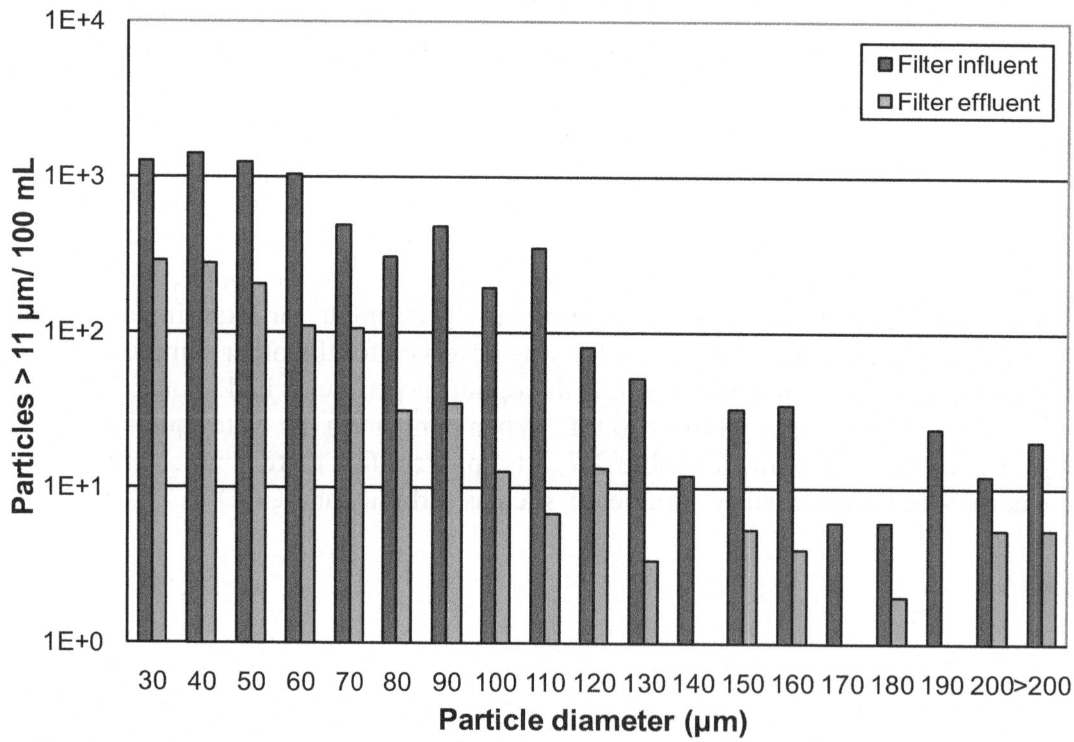


Figure 3-9. Particle Size Distribution Before and After a Cloth Disk Filtration System, Facility A.

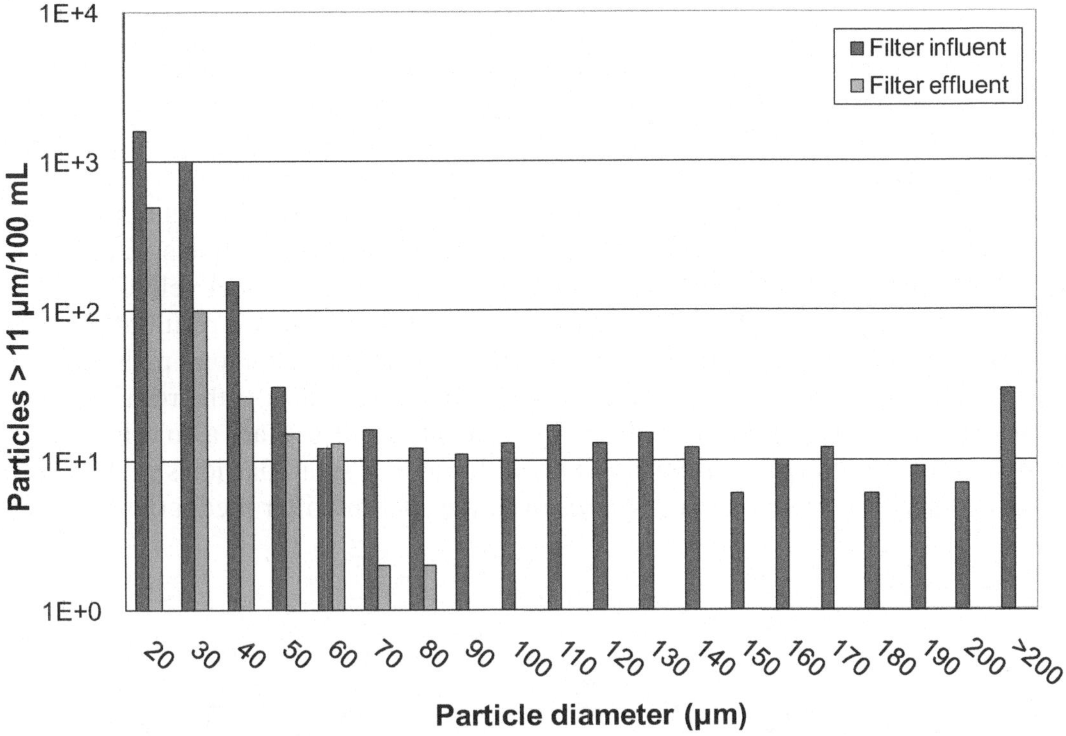


**Figure 3-10. Particle Size Distribution Before and After a Deep-Bed Gravity Filtration System, Facility B.**

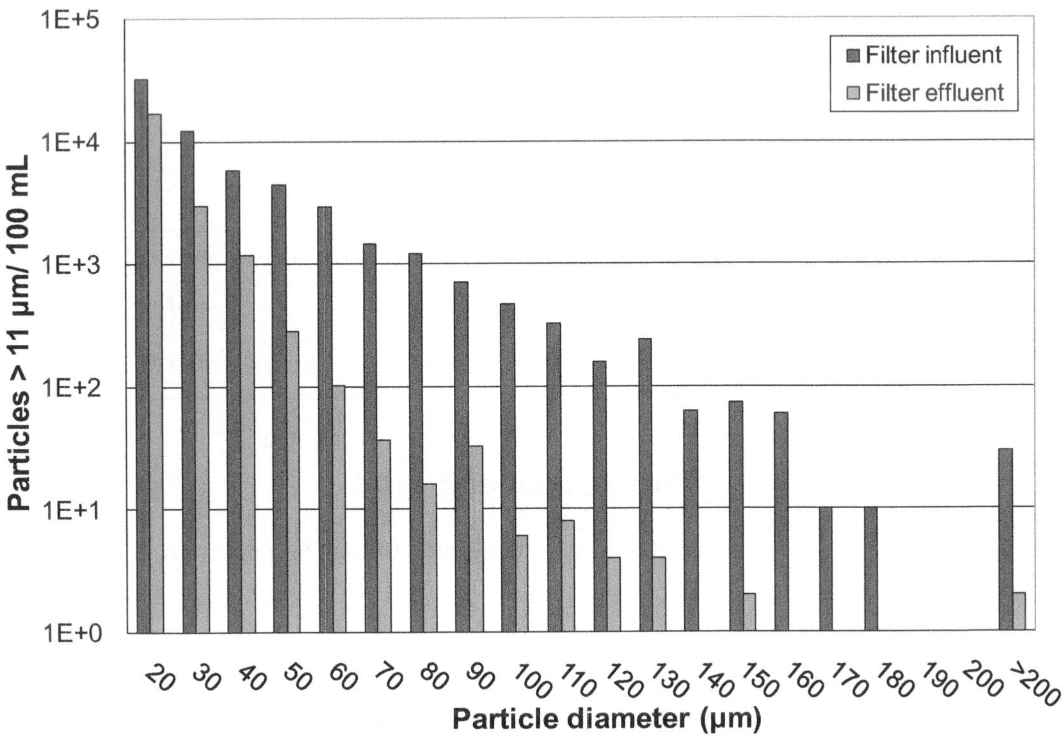


**Figure 3-11. Particle Size Distribution Before and After a Pulsed-Bed Sand Filtration System, Facility F.**

All filters removed a substantial quantity of particles in all particle size classes, although there was preferential removal of larger particles (larger than 100 m), as is shown in Figure 3-12. Percent removal rates for Facility A were variable, especially for the larger particle size classes, but much of the variability can be attributed to low concentrations and variability due to the grab sample method and due to uncertainty in the particle counting method. According to plant operators at Facility A, re-formation of particles occurs in the effluent channels and pipes that are conservatively designed for future flow conditions, resulting in low channel velocities. The removal rates at Facility B are high for large particles, which is probably a result of the deep-bed gravity filtration system used that causes process water to flow through a long pathway through the filter media. The dip in percent removal rates for the Facility B data is the result of very low particle counts for a specific particle size class. The filters at Facility F are also effective at removing large particles although there is more breakthrough of small particles in the Facility F filter than in the other filters, which may be a result of the shallow filter media (less than one foot thick).

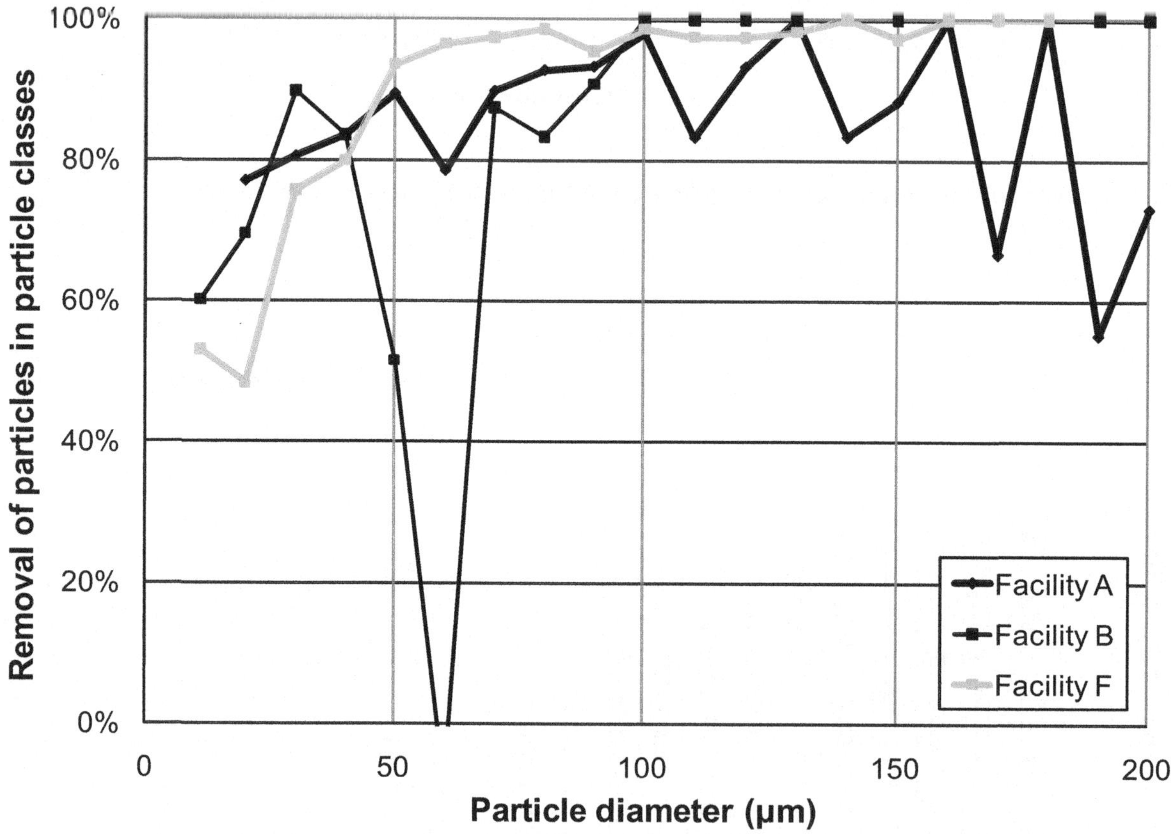


**Figure 3-12. Percent Removal of Particles in Individual Particle Classes for Facilities A, B, and F.**

## 3.7 Discussion and Conclusions

Samples were collected from seven facilities with variable treatment trains, operating process control parameters, and water quality conditions. The samples were used in a series of laboratory tests to quantify disinfection and particle characteristics, which were necessary for model calibration. As a result of the variability in processes and control parameters used at the seven facilities, disinfection and particle characteristics were variable as well.

Information provided by the facilities demonstrated the variability of operating conditions. For example, the MCRT was less than five days at three of the facilities, resulting in non-nitrified effluents. Two of the facilities reported high MCRT values (20 days and 40 days) that resulted in nitrified effluents, which caused very low effluent ammonia concentrations. The two remaining facilities had MCRT values of approximately five to 10 days, which caused the effluent to have a small amount of ammonia (approximately 1 mg/L), and these facilities were considered partially nitrifying in this study. Samples collected from the facilities with nitrified effluents had ammonia nitrogen concentrations below the detection limits (0.02 mg/L) and samples collected from facilities with non-nitrified effluents had ammonia nitrogen concentrations above 20 mg/L.

Water quality data collected in this study (turbidity, UV absorbance at 254 nm, pH, COD, ammonia nitrogen, wet density of particles, and PSD) further demonstrated the variability amongst the facilities. Turbidity measurements were lower for the nitrified effluent samples than for the partially-nitrified and non-nitrified samples. The UV absorbance at 254 nm was fairly consistent for all of the samples except for the sample from Facility A, which had a much lower UV absorbance value, and the sample from Facility G had a UV absorbance value that was much higher. The pH of all samples was consistently between 7 and 8. Consideration of pH is important for the nitrified effluents since the speciation of free chlorine residual products is highly pH-dependent. For example the sample collected from Facility A had a pH of 8.0, which indicated a higher proportion of hypochlorite ions ($OCl^-$) relative to hypochlorous acid (HOCl) than for the sample from Facility B, which had a pH of 7.2. Since $OCl^-$ is a less effective disinfectant than HOCl, additional chlorine would be expected for the sample from Facility A to achieve the same level of inactivation. COD data were collected in hopes of establishing a correlation between COD and some of the model parameters (Chapter 4.0). The COD value observed for Facility A was very low, 3 mg/L, while the values for Facilities D and B were moderate, 18 and 19 mg/L, respectively. The COD values for the other facilities ranged from 30 to 40 mg/L.

A series of disinfection tests was performed to observe the inactivation of total coliform bacteria following the introduction of chlorine into bench-scale reactors. Initial chlorine concentrations were selected that resulted in measurable concentrations of total coliform bacteria in the tailing region of the inactivation curves. Some of the disinfection tests were performed using filtered (2 μm) samples to determine the inactivation rate coefficient of dispersed microorganisms, which was assumed equivalent to the inactivation rate coefficient for occluded microorganisms. The first-order kinetics model worked reasonably well for modeling disinfection data collected using filtered secondary effluent samples. Inactivation curves for disinfection tests performed using filtered (2 μm) samples generally did not have tailing, indicating that the filtration method was effective and that tailing observed in disinfection tests performed using whole water samples was primarily the result of particles. The absence of tailing in disinfection tests performed using filtered water samples also suggests that the chlorine added to the reactors was sufficient to inactivate the microorganisms present. Adding an insufficient amount of chlorine did not appear to be causing tailing in disinfection tests performed using whole water samples because a measurable chlorine residual was present at the end of the disinfection tests.

The inactivation rate coefficients for dispersed (passing through a 2 μm filter) total coliform bacteria varied for the seven facilities. The inactivation rate coefficients for the facilities with the non-nitrified effluents varied from 0.094 to 0.23 L/mg*min. The inactivation rate coefficients for the facilities with the partially-nitrified effluents were 0.17 and 0.55 L/mg*min, which were similar to the values observed for the non-nitrified effluents. The inactivation rate coefficients for the nitrified effluents, where free chlorine was monitored instead of total chlorine, were very different: 8.2 and 14 L/mg*min. As a comparison, Olivieri (1985) reported inactivation data for *E. coli* that was used in Equation 1-2 to produce the following inactivation rate coefficients: 115 L/mg*min when hypochlorous acid was the disinfectant, 5 L/mg*min when hypochlorite was the disinfectant, 0.072 L/mg*min when monochloramine was the disinfectant, and 0.84 L/mg*min when dichloramine was the disinfectant. In general, the inactivation rate coefficient values observed in this study are in agreement with the values reported by Olivieri (1985). There are variations that are a result of the data that are presented by Olivieri (1985); in the disinfection tests the pH conditions were maintained at extreme values (e.g., 4.5 and 10) to isolate the effects of particular chlorine species (e.g., dichloramine). In this study the original pH of the samples was maintained and the disinfectant is a combination of chlorine species as a result (e.g., hypochlorous acid and hypochlorite compose the free chlorine residual in the tests where nitrified effluents were used). In contrast to the data reported by Olivieri (1985), Rice et al. (1999) reported data regarding the inactivation of *E. coli* O157:H7 using chlorine in the absence of ammonia; when Equation 1-2 was used to express their data, the resulting inactivation rate coefficient was 0.89 L/mg*min. One explanation for the difference in inactivation rates is that the disinfection tests by Rice et al. (1999) were performed using free chlorine in demand-free aqueous solutions. As expected, total coliform bacteria disinfected with free chlorine in this study were much more susceptible to inactivation than total coliform bacteria disinfected with combined chlorine. The values of inactivation rate coefficients observed in this study are possibly impacted by the pore size of the filters used to remove particles from the samples. The small pore size of the filter used in this study (2 μm) may have preferentially selected for small and injured bacteria, thereby causing the observed inactivation rate coefficients to be higher than if filters with larger pore sizes had been used.

The chlorine demand was variable for secondary effluent samples collected from the seven facilities, making it difficult to select an initial concentration of chlorine that resulted in an inactivation curve with a substantial concentration of residual total coliform bacteria. The presence of a tailing region in the inactivation curves is necessary to provide sufficient data for model calibration although the data collected from Facility G had a high tailing region. The experiment was completed twice using samples from Facility G with different initial chlorine concentrations. Subsequent experiments would be necessary to obtained disinfection tests with lower tailing residuals of total coliform bacteria. Even for samples collected from the same facility during the validation study, there was a range in optimal chlorine concentrations. It was often necessary to conduct several disinfection tests for a single facility to identify the most appropriate initial chlorine concentration. Adding too little chlorine resulted in insufficient inactivation and adding too much chlorine resulted in concentrations of microorganisms that were below detection limits.

Generation of an appropriate inactivation curve, containing an unambiguous tailing region, was further complicated by the low concentrations of total coliform bacteria in some of the secondary effluent samples. The total coliform concentrations differed by several orders of magnitude for samples collected from the seven facilities. In this study it appeared that low concentrations of target microorganisms were often accompanied by low concentrations of occluded target microorganisms, making it difficult to perform a disinfection test with a measurable tailing region. In future optimization efforts, it would be possible to use an alternate indicator organism although investigators would need to verify that tailing was attributed to microbial occlusion in particles. For examples, it is difficult to use heterotrophic bacteria because this broad category of microorganisms includes a subpopulation that is resistant to inactivation with chlorine, and tailing is evident in inactivation curves produced using data from chlorine disinfection tests as a result.

To determine the frequency of association (F) for secondary effluent samples collected from the facilities, samples were first irradiated with a low dose of UV light and then subjected to the total coliform test. This method yielded low results for F in most cases, typically lower than the range of 0.44 to 24% that was reported by Loge et al. (2002). It is possible that irradiation with a low dose of UV light inactivated a fraction of the occluded microorganisms, especially the smaller particles since particle size is known to effect UV disinfection efficacy (Emerick et al., 1999). Estimates for F are needed for calculation of the particle guidelines, and low estimates of F result in particles guidelines that are too high and not sufficiently conservative. Using the more conservative published values or a value of one may represent a better option for estimating F, rather than using the measured values.

The low frequency of association values might appear to suggest that microbial occlusion in particles is not a significant problem in wastewater effluents. However, microorganisms can partition into the largest particles, that are the most difficult to penetrate with disinfectants, and can be present in the most remote part of the largest particles (e.g. the particle center). If occluded target microorganisms are present in concentrations that are less than the treatment objective, it is not necessary to remove particles or design and operate disinfection systems to inactivate occluded microorganisms. Where stringent discharge standards are in place and highly treated effluents are desired, treatment facilities must be designed and operated to sufficiently inactivate occluded microorganisms. For the data collected in this study, the initial concentration of occluded total coliform bacteria was always much greater than 2 MPN/100 mL (the detection limit for the total coliform test), as was evident from the presence of tailing at concentrations above the detection limits in the disinfection tests. To determine the initial concentration of occluded total coliform bacteria in the tests, it was more efficient to view the results from the disinfection tests and to extend a best-fit line through the tailing region of the inactivation curve. By extending the best-fit line to the y-axis (representing a disinfectant dose of zero), the initial concentration of occluded total coliforms could be estimated. It was necessary to identify the initial concentration of occluded total coliform because it is a model input.

The particle porosities measured for secondary effluent particles collected from the seven facilities varied from 84 to 91%. In future efforts, a sensitivity analysis for this variable might be warranted to determine the impact of its variance on the modeling process. Given the small range of values that were observed, it might be advisable to eliminate this test from the optimization protocols. The particle wet density test is resource-intensive and it may be preferable to devote more resources to collecting additional disinfection test data. In the model simulation exercise the original wet density of particles value was used and the test was not repeated.

Particle data collected from all seven facilities corroborates the claim that there are many particles in secondary effluents and that many of these particles are large (e.g. larger than 20 μm in diameter). Total particles counts for particles larger than 11 μm varied by two orders of magnitude for samples collected at different facilities. High concentrations of particles (diameters greater than 11 µm) were observed in all secondary effluent samples. The total particle counts were much lower for the nitrified effluent samples (3000 particles per 100 mL for Facility B and 4900 particles per 100 mL for Facility A) than for the other samples. Very high particle counts were observed for two of the samples: 150,000 particles per 100 mL for Facility G and 160,000 particles per 100 mL for Facility C. The other particle counts ranged from approximately 60,000 to 82,000 particles per 100 mL. The distribution of particles for the different particle classes also varied for samples collected at the various facilities, although most particles were small (less than 20 µm). Turbidity measurements did not necessary correlate with total particle counts, demonstrating its ineffectiveness as an indicator of total particle presence.

A disadvantage of the microscope imaging method used to collect PSD data is that interpretation of the images is required. For example, the filter mesh can distort particle shape and size (particularly the 80 µm filter) because the particles can become wrapped around the filter material. Also, some of the secondary effluent samples contained algae and other environmental debris and these materials were not included in particle counts. Although the microscope particle counting method resulted in more detailed information, it was time-consuming process. Automated particle counters are an easier way of procuring particle data, however, the accuracy of these instruments is often questioned, especially for environmental samples.

Although aerobic endospores, enterococci, total coliforms, fecal coliforms, *E. coli*, and heterotrophic bacteria were reliably found in secondary effluent samples collected from Facility A, indigenous coliphages were not detected. Results for total coliforms, fecal coliforms, and *E. coli* were nearly the same, which was unexpected because one would expect total coliform concentrations to be higher since fecal coliform and *E. coli* are subsets of the broad group of total coliform bacteria. It is possible that fecal coliform bacteria would have been present in concentrations lower than those of total coliform bacteria had the samples been collected following disinfection. One possible explanation for the high concentration of *E. coli* observed is that it was enumerated using a plate count assay instead of a multiple-tube test and the differences could be attributed to the difference in the type of test used. Heterotrophic bacteria appeared to be the best candidate for use in particle disruption tests because heterotrophic bacteria were abundantly present in secondary effluent samples, including secondary effluent samples that had undergone low doses of UV irradiation. The frequency of aerobic endospores in secondary effluent particles did not appear high enough to make them a good candidate for use in particle disruption experiments as demonstrated by the absence of an increase in aerobic endospores following particle disruption. Additionally, aerobic endospores may not be ideal for use in particle disruption experiments as a result of the detection method procedure, which includes heating samples to 75-80°C prior to the membrane filtration. Heating samples likely changes the particle properties and may skew the results of the particle disruption experiment. For example, heating samples may cause particles to aggregate.

The particle disruption method of Parker and Darby (1995) proved more effective than the method described by Caron et al. (2007) for the release of heterotrophic bacteria from wastewater particles. In the work by Parker and Darby (1995), the ratio of total coliform bacteria in secondary effluent samples that underwent irradiation and particle disruption relative to the total coliform bacteria in irradiated samples ranged from 0.9 to 7.7. A similar ratio obtained in this study for heterotrophic bacteria was 93. It is likely that a larger number of heterotrophic bacteria may have been released from wastewater particles if additional resources had been dedicated to optimizing the particle disruption method for samples collected from Facility A. Neither of the particle disruption methods was effective for increasing the concentration of aerobic endospores.

There are challenges inherent in particle disruption experiments. Previous attempts have met with some success, but most researchers acknowledge that the microbial counts from disrupted samples only represent a fraction of the total target microorganisms present in the particles. Optimal particle disruption strategies are most likely site-specific and vary depending on the target microorganism, as the wastewater particle matrix is highly complex and variable. For example, the surface charge of wastewater particles is variable and probably alters the effectiveness of chemical additions, which constitute a component of particle disruption strategies.

There is uncertainty regarding the optimal UV dose to use to irradiate samples such that only the dispersed fraction is inactivated and the occluded fraction is preserved. UV disinfection data for indigenous aerobic endospores in wastewater were not identified, although it is probably best to perform experiments using samples collected from the facility evaluated in this study. The results presented here show that a substantial portion of aerobic endospores were inactivated using a UV dose of approximately 70 $mJ/cm^2$ (1-2 log removal). The results for heterotrophic bacteria indicate that the UV dose of 35 $mJ/cm^2$ may be insufficient to isolate heterotrophic bacteria that are occluded in particles larger than 11 μm. The assay used for heterotrophic bacteria in this study targets a diverse group of microorganisms that respond differently to UV disinfection. For example, aerobic bacteria that form endospores are a subset of heterotrophic bacteria and are resistant to inactivation with UV light; however, the concentration of aerobic endospores is much lower than the concentration of heterotrophic bacteria (more than one order of magnitude difference). It would be beneficial to produce an inactivation curves for each target microorganism using secondary effluent samples to determine the UV doses that correspond with the beginning of the tailing region in the inactivation curve prior to conducting future particle disruption experiments.

In addition to the data collected from the seven facilities to calibrate the radial pore diffusion model, three additional sets of data were collected from samples collected at Facility A to validation use of the model. Water quality data were fairly consistent for all of the samples collected in the validation study, although there was variation in the PSD data collected. Total particle counts (for particles larger than 11 μm) for the second two samples were much higher than for the first two samples. However, the second two samples had much higher concentrations of the smallest particles, suggesting that although the total particle counts were higher there were fewer large particles that would adversely impact disinfection. Conducting multiple disinfection tests for one facility gave some indication of the temporal variability in chlorine demand although all of the samples were collected during the dry season. Although 8 mg/L was an effective initial chlorine concentration in the first disinfection test conducted for Facility A, it was too high for subsequent tests and lower values yielded much more substantial tailing concentrations of total coliform bacteria. The higher concentrations of small particles in the second two experiments did not appear to adversely impact disinfection, as lower concentrations of chlorine were required to achieve similar levels of inactivation.

Where available, PSD data were collected before and after filtration systems. In general, filtration systems removed a majority of particles in most particle size classes; however, a substantial concentration of particles persisted in the filtered effluents. The purpose of collecting pre- and post- filtration data was to determine if the assumptions made during model simulations (e.g. 30% removal of particles in all particle size classes) were reasonable. Filtration systems removed large particles more effectively than small particles; most particles larger than 100 μm were removed. There was a trend between particle size and percent removal for particles with diameters less than 100 μm. Removal of particles in filtration systems is limited by the capability of existing technologies and removal efficiencies vary. Often, filtration performance is gauged in reference to multiple data sets collected from a single facility (benchmarking).

A limitation of this work is that each set of experiments requires considerable resources, limiting the total number of experiments and replicates that can be performed. The experimental methods result in data with inherent variability. For example, the culture-based microbiological assays have large confidence intervals although these confidence intervals could be narrowed if more replications were incorporated into the experimental design. Increasing the number of replicates would be feasible if more resources were devoted to a single facility rather than trying to conduct the protocols at a variety of facilities. One way to address uncertainty in the experimental methods is to repeat the experimental protocols at a single facility multiple times, calibrate the model separately for each set of experiments, and develop a range of values for the model parameters and for the estimates of intra-particle chlorine concentrations.

CHAPTER 4.0

# USE OF A REACTIVE RADIAL PORE DIFFUSION MODEL FOR INTRA-PARTICLE INACTIVATION

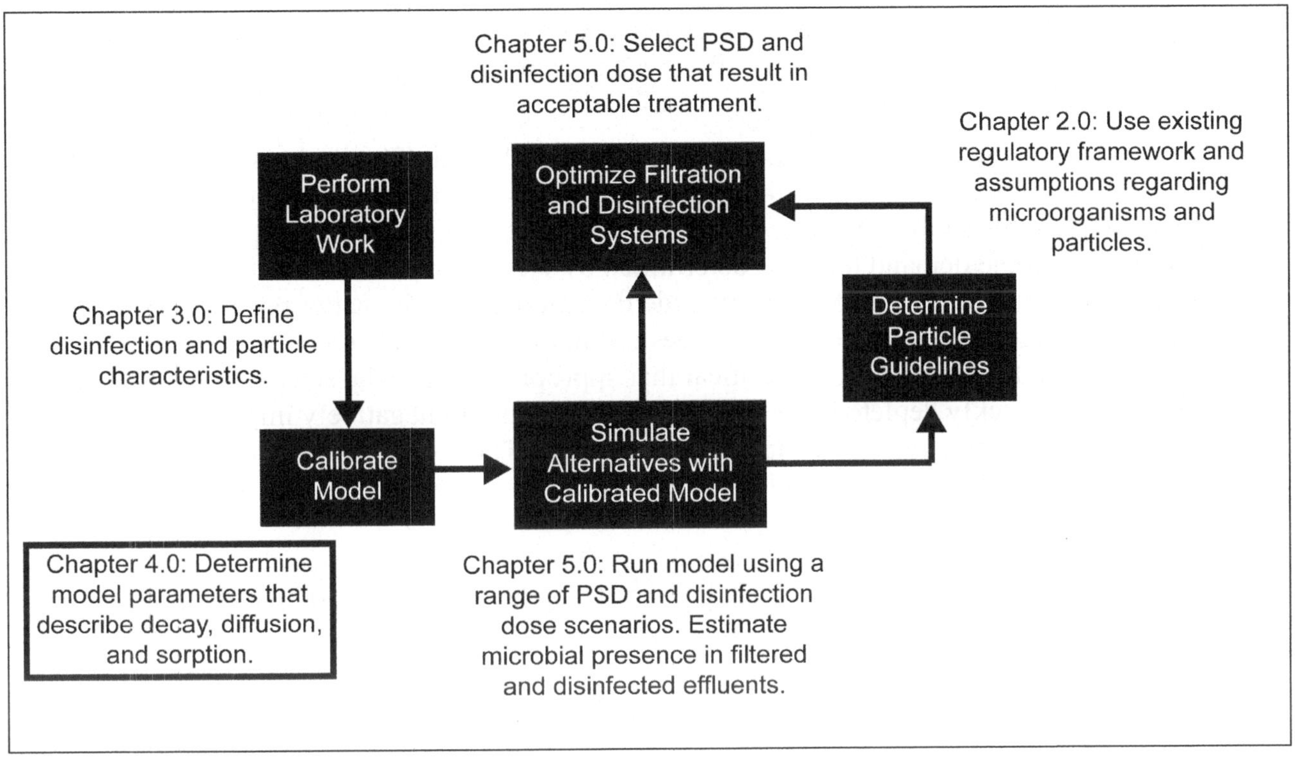


## 4.1 Introduction

A radial pore diffusion model was calibrated independently for each of the seven facilities using data collected in the laboratory studies. Included in this chapter is a general description of the model and its foundations, with citations that provide links to additional information. Model inputs and outputs are described. A summary of data gleaned from the experiments for model calibration (described in Chapter 3.0) is provided. In the final section of this chapter a portion of the model output is shown for the different particle sizes at various time intervals in the disinfection tests. In addition to calculating intra-particle chlorine concentrations, the model calculates the concentration of microorganisms occluded in the center of particles and updates this value for each time step that can be compared with the treatment objectives that were used to calculate the particle guidelines (Chapters 2.0 and 5.0).

## 4.2 Fundamental Bases for the Radial Pore Diffusion Model

### 4.2.1 Core Equations for the Model

A reactive radial pore diffusion model was used to predict intra-particle chlorine concentrations following calibration with experimental data. The model was developed by Basagaoglu et al. (2002) and subsequently used by Dietrich et al. (2003) and Dietrich et al. (2007) to model the penetration of chlorine into wastewater particles. The model allows examination of the mass transfer of a dissolved constituent from the bulk aqueous phase to the interiors of spherical porous particles. Four particles sizes are considered and the intra-particle aqueous and sorbed concentrations are calculated at various radial distances within the particles for a time series. First-order decay of the dissolved constituent occurs in both the bulk fluid exterior to the particles and in the interior of particles.

For the studies by Dietrich et al. (2003) and Dietrich et al. (2007), the model was altered to account for non-linear reactions in the bulk solution that result from reactions with soluble organic matter. In this study, however, it was deemed more efficient to manually account for the instantaneous chlorine demand that was determined by viewing the data collected in the disinfection tests. It would be difficult to include the soluble organic matter parameters in an automated calibration procedure, which was desired in this study. When using the soluble organic matter model parameters, it is critical that appropriate values be selected such that the organic matter is quickly depleted, but care must be taken not to negatively impact the mass balance error by selecting rate constants that are too high. The model incorporates the following equations using dimensionless parameters (nomenclature is described in Table 4-1):

$$\frac{\partial C}{\partial \tau}+\frac{\partial S}{\partial \tau}=\frac{1}{\xi^2}\frac{\partial}{\partial \xi}\left(\xi^2\frac{\partial C}{\partial \xi}\right)-9\varphi^2 C \qquad \tau>0,\ 0\leq\xi\leq 1 \qquad (4\text{-}1)$$

$$\frac{\partial S}{\partial \tau}=\eta Bi(K_d C-S) \qquad \tau>0,\ 0\leq\xi\leq 1 \qquad (4\text{-}2)$$

$$\frac{\partial C}{\partial \xi}=Bi\left(C_{bulk}-K_p C(\xi=1)\right) \qquad \text{at } \xi=1 \qquad (4\text{-}3)$$

$$\frac{\partial C}{\partial \xi}=0 \qquad \text{at } \xi=0 \qquad (4\text{-}4)$$

$$C(\xi,\tau=0)=C_0 \qquad S(\xi,\tau=0)=S_0 \qquad C_{bulk}(\xi,\tau=0)=C_{bulk,0} \qquad (4\text{-}5)$$

$$\frac{\partial C_{bulk}}{\partial \tau}=-3\varepsilon Bi\left(C_{bulk}-C(\xi=1)\right)-9\left((\varphi')^2+(\varphi_{sm})^2 C_{sm}\right)C_{bulk} \qquad (4\text{-}6)$$

$$\frac{\partial C_{SM}}{\partial \tau}=-9(\varphi_{sm})^2 C_{sm} C_{bulk} \qquad (4\text{-}7)$$

**Table 4-1. Model Nomenclature[1].**

| Notation | Description |
|---|---|
| $A$ | surface area of a spherical particle, $A=4\pi R^2$ ($M^2$) |
| $Bi$ | mass transfer associated Biot number, $Bi=Rk_m/\theta D_a$ |
| $C$ | dimensionless intra-particle aqueous concentration, $C=c/c_{ref}$ |
| $C_{bulk}$ | dimensionless bulk fluid concentration, $C_{bulk}=c_{bulk}/c_{ref}$ |
| $C_{in}$ | dimensionless total intra-particle concentration, $C_{in}=C+S$ |
| $C_{sm}$ | dimensionless soluble organic matter concentration in the bulk fluid, $C_{sm}=c_{sm}/c_{ref}$ |
| $D_a$ | intra-particle diffusion coefficient ($L^2/T$), $D_a=D^*\theta^{1/3}$ |
| $D$ | bulk fluid diffusion coefficient ($L^2/T$) |
| $K_d$ | ratio of the sorption rate to the desorption rate, $K_d=k_f/k_r$ |
| $K_p$ | equilibrium partition coefficient at particle surface |
| $R$ | reference particle radius (L) |
| $R_i$ | dimensionless particle radius |
| $S$ | dimensionless intra-particle sorbed concentration, $S=(\rho/\theta)(s/c_{ref})$ |
| $V$ | volume of the bulk fluid outside of the particles ($L^3$) |
| | |
| $c$ | intra-particle aqueous concentration ($M/L^3$) |
| $c_{bulk}$ | bulk fluid aqueous concentration ($M/L^3$) |
| $c_{ref}$ | maximum concentration observed initially either in the reacted or dissolved phase, $c_{ref}=max_{\tau=0}\{C,S,C_{bulk}\}$, ($M/L^3$) |
| $c_{sm}$ | soluble organic in the bulk fluid concentration ($M/L^3$) |
| $k$ | intra-particle aqueous decay rate ($T^{-1}$) |
| $k'$ | bulk fluid aqueous decay rate ($T^{-1}$) |
| $k_f$ | surface oxidation reaction coefficient ($T^{-1}$) |
| $k_i$ | inactivation rate coefficient ($L^3/(M^*T)$) |
| $k_m$ | mass-transfer coefficient across the particle boundary (L/T) |
| $k_r$ | desportion rate coefficient ($T^{-1}$) |
| $k_{sm}$ | soluble organic matter decay rate ($T^{-1}$) |
| $r$ | spatial index along the radius (L), $0 \leq r \leq R$ |
| $s$ | intra-particle sorbed mass per mass of particle (M/M) |
| $t$ | time (T) |
| | |
| $\varepsilon$ | ratio of intra-particle aqueous volume of a particle to extra-particle aqueous volume, $\varepsilon=\Sigma\theta(4/3)\pi R_i^3/V$ |
| $\eta$ | relative importance of intra-particle desorption to mass transfer at the particle surface, $\eta=\theta k_r R/k_m$ |
| $\theta$ | particle porosity, ratio of intra-particle aqueous volume to the total particle volume |
| $\xi$ | dimensionless length, $\xi=r/R$, $\xi=0$ at the particle center and $\xi=1$ at the particle surface |
| $\rho_b$ | Mass of sorbent per total particle volume ($M/L^3$) |
| $\tau$ | dimensionless time, $\tau=tD_a/R^2$ where the $R$ used was the radius of the largest particle size class |
| $\varphi$ | reaction/diffusion Thiele modulus for the intra-particle aqueous phase, $\varphi=(R/3)(k/D_a)^{0.5}$ |
| $\varphi'$ | reaction/diffusion Thiele modulus for bulk fluid, $\varphi'=(R/3)(k'/D_a)^{0.5}$ |
| $\varphi_{sm}$ | reaction/diffusion Thiele modulus for soluble organic matter, $\varphi_{sm}=(R/3)(k_{SM}/D_a)^{0.5}$ |
| $\omega_j$ | number of particles in the jth size class |

[1]Adapted from Basagaoglu et al. 2002, Dietrich et al. 2003, and Dietrich et al. 2007.

Equation 4-1 describes the time rate of change of the disinfectant in both the aqueous and sorbed phases in the interior of the particles. The model allows for reversible sorption to occur inside of particles, and this phenomenon is described by Equation 4-2. Mass transfer at the particle surface is described in Equation 4-3, whereas mass transfer at the particle center is assumed to be zero as shown in Equation 4-4. The initial conditions for the bulk and intra-particle concentrations are described in Equation 4-5. Equation 4-6 describes the time rate of change of the disinfectant concentration in the bulk solution. Equation 4-7 describes the time rate of change of the concentration of reactive organic matter in the bulk solution. For a more complete description of the model functionality, refer to Basagaoglu et al. (2002).

### 4.2.2 Conservation of Mass Inside of Particles

The mass conservation of chlorine inside of particles is described by Equation 4-1, which is shown in original dimensional form in Equation 4-8. The total change in aqueous and sorbed chlorine mass inside of particles is balanced by the diffusive mass flux into or out of the particle and the decay of chlorine inside the particle, respectively in both Equations 4-1 and 4-8.

$$\theta\frac{\partial c}{\partial t}+\rho_b\frac{\partial s}{\partial t}=\frac{\theta D_a}{r^2}\frac{\partial}{\partial r}\left(r^2\frac{\partial c}{\partial r}\right)-\theta kc \qquad t>0,\ 0\leq r\leq R \qquad (4\text{-}8)$$

### 4.2.3 Sorption of Chlorine in Particle Pore Spaces

Reactive chlorine species may adsorb onto the exterior particle surface or onto the surfaces in the pore spaces. Because the surface area of the particle pore spaces is thought to be much greater than the exterior surface area of the particles, only adsorption in the pore spaces is considered. The high values of particle porosity reported in Chapter 3.0 support this hypothesis although the measured porosities reflect total particle porosity, which includes the porosity of particle organic matter and occluded bacteria. A first-order kinetics model was assumed for sorption, where the rate of sorption was assumed to be a function of the difference of the sorbed and aqueous concentrations from their equilibrium values. The sorption model was based on the assumption that there are an unlimited number of sorption sites available. Equation 4-12 is the dimensional form of Equation 4-2, which describes the first-order sorption model.

$$\frac{\partial s}{\partial t}=k_r\left(K_d c-s\right) \qquad \tau>0,\ 0\leq\xi\leq 1 \qquad (4\text{-}12)$$

### 4.2.4 Thin Film at the Particle Surface

At the particle surface, the mass transfer of chlorine is described by Equation 4-3, which is the dimensionless form of Equation 4-9. A first-order kinetics model was again assumed for mass transfer between the particle (aqueous) boundary and the bulk fluid phase, across a "thin film" boundary layer. This rate law is consistent with the idea that there is a thin film of stagnant fluid (boundary layer) surrounding the particle. Chlorine constituents must travel through the thin film to reach the pore spaces, and a concentration gradient can exist across the thin film as a result of diffusive mass transfer limitations.

$$\theta D_a\frac{\partial c}{\partial r}=k_m\left(c_{bulk}-K_p c(R)\right) \qquad at\ r=R \qquad (4\text{-}9)$$

### 4.2.5 Fick's First Law of Diffusion for Flux into Particles

To inactivate occluded microorganisms, chlorine disinfectants must first cross the boundary layer and then diffuse through the particle pore spaces to make contact with the occluded microorganisms. Diffusion is described using Fick's First Law (Equation 4-10) that states that the diffusive flux $J$ of a constituent across a finite surface area ($ML^{-2}\ T^{-1}$) is proportional to the gradient of concentration; $D$ is the diffusion coefficient ($L^2\ T^{-1}$), The diffusion coefficient ($D$) described by Fick's Law pertains to diffusion of constituents in the bulk fluid while the intra-particle diffusion coefficient ($D_a$) is used in the formulation of the radial pore diffusion model. The diffusion coefficient ($D_a$) is not found in any of the core Equations 4-1 to 4-7 because it was factored out of the equations when the physical equations were transformed into their dimensionless forms (equations for the Biot number ($Bi$), dimensionless time ($\tau$), and the Thieli moduli ($\varphi$, $\varphi'$, and $\varphi_{sm}$) are shown in Table 4-1).

$$J = -D\frac{\partial \phi_f}{\partial x} \qquad (4\text{-}10)$$

The diffusive flux at the particle surface is described with Equation 4-11, which is based on Fick's First Law. Mass transfer from the bulk fluid into the particles is assumed to occur at the particle surface without any losses (decay). One model assumption is that the bulk fluid is well-mixed and that localized areas of concentrated chlorine species are not existent. Following conversion of the dimensional equation into its non-dimensional form, Equation 4-11 was transformed to Equation 4-6 with the addition of a term to account for chlorine demand from the soluble organic matter as well.

$$V\frac{\partial c_{bulk}}{\partial t} = -\omega AD\frac{\partial c}{\partial r}\bigg|_{r=R} = -\omega Ak_m\left(c_{bulk} - K_p c(R)\right) - Vk'c_{bulk} \qquad (4\text{-}11)$$

### 4.2.6 First-Order Decay of Reactive Chlorine Species

The reactive chlorine species tracked by the model are assumed to have first-order decay kinetics, which are inherent in the core dimensionless Equations 4-1 to 4-7. The first-order decay coefficients ($k$, $k'$, and $k_{sm}$) are represented by the model parameters $\varphi$, $\varphi'$, and $\varphi_{sm}$. Decay of chlorine species can occur in the bulk fluid or inside of particles; in the bulk fluid chlorine decay can occur as a result of reactions with the soluble organic matter or it can occur as the result of other phenomena. In the dimensional Equation 4-8 the first-order relationship is more apparent, and the rate of change of the aqueous chlorine concentration is a function of the concentration. In Equation 4-8 the first-order decay term is indicated for intra-particle chlorine concentrations. A similar relationship is shown in Equation 4-11 for first-order decay of chlorine in the bulk fluid. There are various reasons for decay of aqueous chlorine species. For example, the presence of algae in the bulk fluid increases chlorine demand, and demand causes reactive chlorine in the bulk solution to decrease or decay.

### 4.2.7 Reaction of Chlorine with Soluble Organic Matter

When chlorine is added to an environmental water sample, a rapid decrease in the bulk fluid chlorine concentration is often observed that occurs when chlorine reacts with the soluble organic matter and metals. In the model the concentration of soluble organic matter that is available to react with chlorine is identified as well as a kinetic parameter. The chlorine demand that occurs as a result of adsorption of reactive chlorine species onto the exterior surfaces of particles is also accounted for in the instantaneous chlorine demand. For this study the instantaneous chlorine demand was accounted for manually and the soluble organic matter model parameters were not used (see Section 4.2.1).

### 4.2.8 Chlorine Chemistry in the Model

The radial pore diffusion model is formulated to evaluate a single reactive species. In this study the reactive species is total chlorine residual and no differentiation is made between the different reactive species that comprise the chlorine residual. Where the model is used for nitrified effluents, no differentiation is made between hypochlorous acid (HOCl) and hypochlorite ($OCl^-$). To prevent changes in the relative proportions of the different chemical species that make up the free chlorine residual, the pH and temperature were maintained during the disinfection tests. Where the model is used for non-nitrified effluents, no differentiation is made between monochloramine ($NH_2Cl$), dichloramine ($NHCl_2$), and other forms of combined chlorine. An assumption used in the model is that the reactivity of chlorine inside of particles is similar to that in the bulk solution, and that the relative proportions of chlorine species are not changing inside of particles.

## 4.3 Model Inputs and Assumptions

The model is empirical and is run and calibrated using experimental data. In this research a procedure was developed to streamline model calibration and facilitate comparison of model parameters selected for the seven facilities; the procedure was based on several assumptions. The initial intra-particle chlorine concentrations (sorbed and aqueous) were assumed zero at the beginning of the disinfection tests. The initial disinfectant concentration in the bulk solution was determined by analyzing the best-fit for chlorine residual data over time. The actual initial chlorine concentration in solution was not used because the instantaneous chlorine demand was assumed to occur prior to transfer of chlorine from the bulk fluid into particles. The instantaneous chlorine demand is a function of the soluble mineral and organic content of the wastewater that is immediately oxidized by the addition of chlorine. Therefore, the initial chlorine concentration, in the context of model calibration, is the applied chlorine concentration minus the instantaneous chlorine demand.

Laboratory data used to run the model consists of the PSD (both $R_i$ and $\omega_i$ data), initial bulk fluid chlorine concentration ($c_{bulk,t=0}$),particle porosity that was determined from the particle wet density data, inactivation rate coefficient for dispersed total coliform bacteria, and the initial concentration of occluded total coliform bacteria that was determined from the results of the disinfection tests ($N_0$). The initial chlorine concentration was defined as the initial concentration of chlorine added to the reactor minus the instantaneous chlorine demand, which was determined by analyzing the measured chlorine residual data. The PSD data were condensed into four particle size classes and the median values for the radius data were used to define $R_i$. In addition to laboratory data, the following model parameters were calculated separately for each of the four particle sizes and the corresponding radius values were used in the model calculations: ratio of intra-particle aqueous volume of a particle to the extra-particle aqueous volume ($\varepsilon_i$), ratio of intra-particle desorption to mass-transfer at the particle surface ($\eta$), dimensionless length ($\xi$), and the reaction and diffusion Thiele moduli inside of particles ($\varphi$) and in the bulk fluid ($\varphi'$).

An attempt was made to calculate *Bi* using calculations for the mass transfer coefficient across the particle boundary ($k_m$), as determined using the data and procedure outlined by Harriott (1962). In the work of Harriott (1962) $k_m$ varies according to particle size and larger particles have smaller mass transfer coefficients. The Biot numbers calculated using $k_m$ values that vary according to particle size will vary as well for different particle sizes. However, in preliminary model calibration efforts using the $k_m$ values of Harriott (1962) the Biot numbers appeared to be too high (ranging from approximately one to 20) and the mass transfer of chlorine across the particle was too fast, resulting in very low chlorine decay rates ($k$ and $k'$) and very high partitioning coefficients ($K_d$). The Biot numbers were subsequently adjusted to 0.1, the lower end of the range identified by researchers where there is mass transfer resistance at the particle surface as well as resistance to mass transfer within the particle that results in an internal gradient (Hsu, 2003). In systems where the Biot number is less than 0.1 the internal resistance is negligible and in systems where the Biot number is greater than 40 there is negligible surface resistance (Hsu, 2003). The Biot number in this study was set to the lower end of the range with finite surface and internal resistance to maximize the effects of decay and minimize the effects of intra-particle sorption.

Diffusion coefficients for free and combined chlorine residual species were needed to calculate $\tau$, $\varphi$, and $\varphi'$. The diffusion coefficients for hypochlorous acid and hypochlorite are similar and an intermediate value of $1.1 \times 10^{-9}$ $m^2/s$ was selected for data collected from facilities with nitrified effluents (Chao, 1968). A diffusion coefficient of $1.9 \times 10^{-9}$ $m^2/s$ was selected for monochloramine and was used for data collected from facilities with non-nitrified effluents (Stewart and Raquepas, 1995). The intra-particle diffusion coefficient was calculated and is a function of the bulk fluid coefficient (Table 4-1).

Although based on dimensional mathematical relationships, the radial pore diffusion model has a dimensionless formulation. To convert dimensionless time ($\tau$) into dimensional time, the following equation is used: $\tau = t^* D_a / R^2$. In the calculation of time the value for R is the largest particle size used in the model, which varies for each of the seven data sets. To achieve model convergence, the dimensionless time step ($\Delta\tau$) was set at 0.01. It was determined to be the largest possible time step based on a sensitivity analysis where the model output did not appreciably change for values smaller than 0.01, but model stability decreased for dimensionless time steps larger than 0.01. The number of dimensionless time steps required for each model run was determined by calculating $\tau$ and dividing this value by the dimensionless time step.

Dimensionless length ($\xi$) is also a function of $R$, although the individual radius values for the four particle size classes are used for its calculation: $\xi_i = r/R_i$. In an analysis completed by Basagaoglu et al. (2002), it was determined that the particles should be discretized into 1001 segments for model convergence to minimize mass balance errors. However, in the work of Basagaoglu et al. (2002) there was very little difference between the model results using 1001 nodes and the results obtained using models with 101 nodes. A sensitivity analysis was performed as part of this study to determine if it would be appropriate to use 101 nodes, which would greatly decrease model computation and runs times. When the model output obtained using 1001 and 101 nodes was compared, there was less than 1% difference in the concentrations of occluded microorganisms and bulk fluid chlorine. The average intra-particle chlorine concentrations were 2.6% different, although the effect on microbial survival at the particle centers was lower. Ultimately, it was determined that 101 nodes would be acceptable for the model runs in this study and $\Delta\xi$ was set at 1/101.

Parameters adjusted during model calibration were $k$, $k'$, and $K_d$. It was assumed that chlorine decay kinetics exterior to and inside of particles are similar and that the rate of decay of chlorine in the bulk fluid ($k'$) is equivalent to the intra-particle chlorine decay rate ($k$), which reduced the number of fitting parameters to three. A parameter estimating software, UCODE, was used to calibrate the model (Poeter et al., 2005). UCODE functions by comparing observations and values calculated by a user-supplied model using a statistical analysis that aims to minimize a least-squares objective function. Nonlinear regression is used and the objective function is minimized using a Gauss-Newton method. Parameter values are sequentially perturbed in an iterative process and the corresponding sensitivities are calculated. The iterative calibration procedure used by UCODE involves first calculating sensitivities for the model parameters that are obtained by perturbing each of the model parameters individually. The UCODE then calculates the best choice of model parameters for the next iteration.

UCODE cannot function well if the model parameters are highly correlated because there is not a unique calibration solution. In this study the Biot number and the ratio $k_r/k_m$ were originally included as model parameters; however, it was not possible to have four model parameters because of highly correlated model parameters. Subsequently, the Biot number was set at 0.1 for all particle sizes and $k_r/k_m$ was set at 0.01 $\mu m^{-1}$. As is apparent in Equation 4-2, $\eta$ and $K_d$ function similarly because they both impact the rate of sorption. If the intra-particle sorbed concentration ($S$) is low, then $\eta$ and $K_d$ will have the same impact on the model results and it is easy to identify the reason that the model parameters are highly correlated. In the formulation of the model in Basagaoglu et al. (2002) η always appears in the numerator and denominator of the equations (Equations 9, 11, 12, 15, 16, 17 in Basagaoglu et al. (2002)) and as a result very large values of $\eta$ will have no effect on the model result. Therefore, $k_r/k_m$ was set at 0.01 $\mu m^{-1}$ so that the values of $\eta$ would be low and the values would not cancel out in the core equations.

Chlorine residual and total coliform concentrations observed during the disinfection test were used to calibrate the model. The total coliform data used were those from the tailing portion of the inactivation curve produced using data from the disinfection test. Typically, tailing started approximately 8 to 10 minutes after the addition of chlorine and the measured total coliform data collected in the tailing region of the disinfection test were used for model calibration. A major assumption of the approach used in this study is that tailing in the disinfection test is the result of occlusion of microorganisms in particles that requires disinfectants to first diffuse through particles prior to inactivating the occluded microorganisms. Many data collected later in the disinfection tests had total coliform concentrations that were below detection limits for the total coliform assay, and only samples that had measurable concentrations of total coliform bacteria could be used for model calibration. The values for total coliform concentration in the tailing region were dependent on the quantity of chlorine added in disinfection tests and on disinfection and particle characteristics for the samples collected.

To obtain better fits for the total coliform data, in particular the low concentrations observed at the end of 90 minutes, the log of the total coliform was used. Both the total coliform observations and the total coliform concentrations calculated by the model were log transformed. Intra-particle chlorine concentrations calculated by the model were used in combination with measured inactivation rate coefficients to determine the concentration of occluded total coliform bacteria (Equation 1-2). It was conservatively assumed that occluded total coliform inhabit the center of particles, and that by ensuring that the centers of particles are adequately saturated with the chlorine that any total coliform bacteria potentially present in the particle are inactivated. The inactivation coefficient used to calculate the concentration of occluded total coliform bacteria was determined experimentally for filtered (2 μm) samples, which is based on the assumption that occluded target microorganisms have the same inactivation characteristics as dispersed target microorganisms and that only diffusion of the chlorine through the particles is limiting inactivation.

To perform the regression analysis based on the sum of the squared residuals (using UCODE), it was necessary to assign weights to both the chlorine and total coliform observations. In this study there was more than one type of observation that was used to calibrate the model. The two types of observations have different units, different accuracies, and there are more of one type of observation than the other. The chlorine data are more accurate and have less variability than the total coliform data that are highly variable. For both types of observations error is introduced in measurement of the samples with higher values because the samples must first be diluted prior to analysis. However, it was desirable to assign higher weights to smaller numbers so that the residual is more significant when compared with the residuals of the observations with higher values. In this study there were typically two to three times as many chlorine data as total coliform data, which automatically resulted in the chlorine data having a stronger influence in the model calibration. In addition, the chlorine data spanned a longer time period; chlorine data were collected for approximately 5-10 hours while the total coliform data were only collected for 90 minutes.

The observations were weighted in two ways: the individual observations were weighted using the coefficient of variation and the two groups of observations (chlorine and total coliform) were assigned a simple weight multiplier. The coefficient of variation is defined as the standard deviation divided by the mean. In UCODE when the coefficient of variation (cv) is used as a weighting method, the weight is calculated as $1/(cv*observation)^2$. For this study cv=0.1 was used for the chlorine data and cv=0.5 was used for the total coliform data, except that cv=0.2 was used for the final total coliform observations to improve the fit of the model for the final total coliform observations. The total coliform data were collected using the MPN method, which is based on a Poisson distribution. A characteristic of the Poisson distribution is that the mean is equal to the variance ($\lambda$). In this study the log (base 10) of the total coliform was used to compare model fits. The coefficient of variation for the total coliform data is 0.5 because $cv=\log10(\lambda^{1/2})/\log10(\lambda)=0.5$.

Initially the chlorine data were assigned much higher weights than the total coliform data because the chlorine data are more accurate and involve fewer assumptions. However, the model fit for the chlorine data was much better than for total coliform data. In subsequent model runs the weight multiplier for the chlorine data was lowered and the effect on the chlorine and total coliform observations was monitored. Appropriate model fits for both chlorine and total coliform were obtained when the weight multipliers for the observation groups were equal. The higher number of chlorine observations provided weighting for the chlorine data that was sufficient to obtain good model fits for the chlorine observations.

## 4.4 Model Calibration Sequence

To calibrate the model the bulk fluid chlorine residual and total coliform concentrations in the tailing region of the disinfection test were used and the following initial values were assigned to the model parameters: $k=k'=0.01$ and $K_d=100$. The model parameters were constrained in the estimating process so that negative values could not be assigned to the parameters. The UCODE model was first run with the optimization and sensitivity functions turned off to verify proper operation of the model. The UCODE model was next run with the sensitivity function turned on to determine the sensitivities of the fitting parameters that affect calculation of the observations. The second step is intended to verify that the process model (the radial pore diffusion model in this case) is functioning properly and that the fitting parameters are correctly assigned. Next, the UCODE model was run in parameter estimation mode where both the optimization and sensitivity functions are turned on. First, the observations were calculated using the starting values of the parameters and the sum of the squared residuals was calculated. Next, the sensitivities were calculated so that new parameters values could be assigned in the next iteration. In the next iteration the parameters were adjusted and a new sum of squared residual value was calculated. The exercise of calculating sensitivities to assign new parameter values was repeated. The iterative process was repeated by the UCODE model until model convergence was achieved. To test that UCODE was indeed selecting an optimal solution and that the solution was unique, different starting values were used to determine if the final result was the same.

## 4.5 Model Calibration for All Facilities

Data used for model calibration are shown in Table 4-2. The PSD data were divided into four particle size classes because the model considers mass transfer in particles with four discrete particle sizes. The total experiment time refers to the time over which chlorine data were collected; total coliform data were only collected for 90 minutes. The initial concentrations of occluded total coliform were determined by viewing the disinfection test data and were then distributed into the four particle size classes. It was assumed that the initial concentration of occluded total coliform bacteria occupied the larger particles prior to the smaller particles. Unfortunately, the total coliform tailing observed in the disinfection test for Facility G was higher than the particle counts for particles larger than 11 μm. The data were insufficient to fully calibrate the model for this facility; however, it was assumed that all particles larger than 11 μm had at least one occluded total coliform bacteria. The assumption for Facility G regarding occluded total coliform bacteria is reasonable because the initial chlorine concentration in the disinfection test was insufficient to inactivate a large portion of the total coliform bacteria observed in the test.

**Table 4-2. Data Used for Model Calibration.**

| | Facility A | Facility B | Facility C | Facility D | Facility E | Facility F | Facility G |
|---|---|---|---|---|---|---|---|
| Particle diameters (μm) | 132<br>84<br>50<br>26 | 296<br>188<br>114<br>18 | 188<br>96<br>48<br>16 | 224<br>162<br>96<br>16 | 248<br>132<br>60<br>16 | 304<br>130<br>76<br>18 | 192<br>118<br>64<br>16 |
| Particle concentrations ($\omega_i$) (#/100 mL) | 20<br>114<br>1806<br>2941 | 14<br>50<br>97<br>2858 | 30<br>710<br>8400<br>150,550 | 13<br>47<br>360<br>80,387 | 6<br>35<br>1727<br>80,256 | 30<br>457<br>4330<br>58,097 | 200<br>350<br>12,567<br>139,033 |
| Total experiment time (min) | 489 | 610 | 495 | 532 | 800 | 532 | 455 |
| Particle porosity ($\theta$) | 0.86 | 0.88 | 0.9 | 0.89 | 0.84 | 0.86 | 0.91 |
| Initial concentration of occluded total coliform ($N_0$) (MPN/100 mL) | 20<br>0<br>0<br>0 | 14<br>20<br>0<br>0 | 17<br>0<br>0<br>0 | 13<br>64<br>0<br>0 | 6<br>35<br>29<br>0 | 22<br>0<br>0<br>0 | 200<br>350<br>12,567<br>139,033 |
| Inactivation rate coefficient ($k_i$) (L/mg*min) | 13.9 | 8.2 | 0.173 | 0.548 | 0.105 | 0.229 | 0.0938 |
| Initial chlorine concentration[1] ($c_{bulk,t=0}$) | 4.9 | 2.9 | 4.0 | 3.4 | 19.6 | 11.2 | 1.0 |

[1]Initial chlorine=chlorine added - instantaneous chlorine demand

The model inputs and the parameters selected during calibration are shown in Table 4-3. The intra-particle diffusion coefficient was calculated using published values for the bulk fluid diffusion coefficient and the particle porosity values. The reference particle is the largest particle size in the model and was used to convert dimensionless time into dimension values. The dimensionless radii were calculated by dividing all dimensional values by the reference radius. Based on separate model sensitivity analyses, the dimensionless time step was set at 0.01 and radii were partitioned into 101 nodes. The Biot number was held constant at 0.1 for all particle sizes and the $k_r/k_m$ ratio used to calculate $\eta$ was set at 100 $cm^{-1}$. Calculated values of $\eta$ ranged from 0.07 to 1.31, depending on the particle size.

**Table 4-3. Model Inputs and Parameters.**

| Model parameter | Facility A | Facility B | Facility C | Facility D | Facility E | Facility F | Facility G |
|---|---|---|---|---|---|---|---|
| Intra-particle diffusion coefficient ($D_a$) ($cm^2$/min) | 0.0006 | 0.0006 | 0.0006 | 0.0006 | 0.0011 | 0.0011 | 0.0011 |
| Reference particle ($R$) (μm) | 132 | 296 | 188 | 224 | 248 | 304 | 192 |
| Dimensionless radius ($R_i$) (μm) | 1.0<br>0.64<br>0.38<br>0.20 | 1.0<br>0.64<br>0.39<br>0.06 | 1.0<br>0.51<br>0.26<br>0.09 | 1.0<br>0.72<br>0.43<br>0.07 | 1.0<br>0.53<br>0.24<br>0.06 | 1.0<br>0.43<br>0.25<br>0.06 | 1.0<br>0.61<br>0.33<br>0.08 |
| Number of nodes | 101 | 101 | 101 | 101 | 101 | 101 | 101 |
| Dimensionless time step ($\Delta\tau$) | 0.01 | 0.01 | 0.01 | 0.01 | 0.01 | 0.01 | 0.01 |
| Number of dimensionless time steps | 674,000 | 168,000 | 337,000 | 255,000 | 573,000 | 254,000 | 544,000 |
| Biot number ($Bi$) | 0.1 | 0.1 | 0.1 | 0.1 | 0.1 | 0.1 | 0.1 |
| Ratio of desorption to mass-transfer across particle boundary ($k_r/k_m$) (1/cm) | 100 | 100 | 100 | 100 | 100 | 100 | 100 |
| Desorption to mass transfer at the particle surface ratio ($\eta$) | 0.57<br>0.36<br>0.22<br>0.11 | 1.30<br>0.83<br>0.50<br>0.08 | 0.85<br>0.43<br>0.22<br>0.07 | 1.0<br>0.72<br>0.43<br>0.07 | 1.04<br>0.55<br>0.25<br>0.07 | 1.31<br>0.56<br>0.33<br>0.08 | 0.87<br>0.54<br>0.29<br>0.07 |
| Ratio of the sorption rate to the desorption rate ($K_d$) | 18,670 | 2465 | 1479 | 2363 | 2656 | 1429 | 1500 (estimated) |
| Decay rates ($k=k'$) (1/min) | 0.0035 | 0.011 | 0.0018 | 0.0014 | 0.00045 | 0.0015 | 0.0054 |
| Thiele modulus, intra-particle ($\varphi$) | 0.005<br>0.003<br>0.002<br>0.001 | 0.021<br>0.013<br>0.008<br>0.001 | 0.005<br>0.003<br>0.001<br>0.0005 | 0.006<br>0.004<br>0.002<br>0.0004 | 0.003<br>0.001<br>0.001<br>0.0002 | 0.006<br>0.003<br>0.001<br>0.0004 | 0.007<br>0.004<br>0.002<br>0.001 |
| Thiele modulus, bulk fluid ($\varphi'$) | 0.005 | 0.021 | 0.005 | 0.006 | 0.003 | 0.006 | 0.007 |

Model calibration yielded values for the model parameters $K_d$ and $k=k'$ for each of the facilities. For Facilities B, C, D, E, and F values of $K_d$ were between 1400 and 2700. The value for Facility A was much higher: 18,670. Since insufficient data were available to calibrate the model for Facility G, a value of $K_d$=1500 was assigned so that the model and simulation exercises could be completed to provide an estimate of results for this facility. The calibration was completed for Facility G to obtain the most appropriate value for $k=k'$. The decay rates for the facilities ranged from 0.00045 to 0.011 min$^{-1}$. The calculated intra-particle Thiele moduli varied from 0.0002 to 0.021 and the bulk fluid Thiele moduli varied from 0.003 to 0.021.

Model output for the bulk fluid chlorine concentration is shown in Figure 4-1 along with experimental data for bulk fluid chlorine residual observed during disinfection tests. The selection of initial chlorine concentrations that provided a best-fit of model output and experimental data is apparent. The shapes of the chlorine residual curves differ for each facility, and are likely a function of concentration of chlorine added and the site-specific disinfection characteristics. Where a high concentration of chlorine was added to reactors, the chlorine decay was more gradual (results for Facility D, E, and F are in Figure 4-1). For the disinfection tests performed using samples from the other four facilities, chlorine residual decay was fairly exponential. In the test for Facility B, the initial chlorine added was insufficient to maintain a chlorine residual throughout the period of time that chlorine was monitored. In the case of Facility B, a low concentration of chlorine was necessary to produce a tailing region in the inactivation curve with measurable concentrations of total coliform bacteria. The challenge of generating data for calibrating the model is to produce a disinfection test that results in a measurable concentration of chlorine in the bulk fluid and a measureable residual concentration of total coliform bacteria. The initial chlorine concentration must be high enough to retain a chlorine residual, but not so high as to completely inactivate the occluded total coliform bacteria.

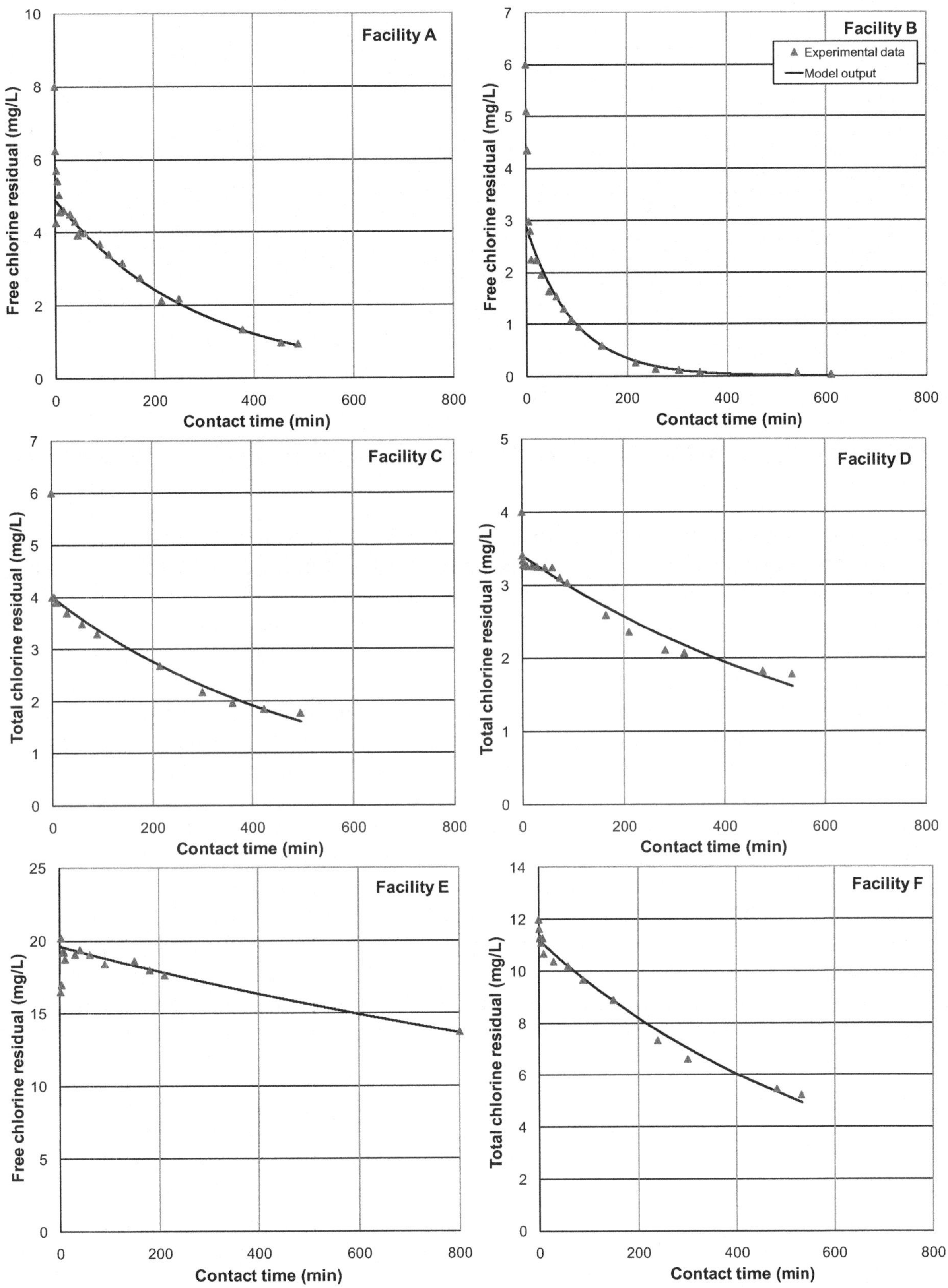


**Figure 4-1. Comparison of Bulk Fluid Chlorine Concentrations Calculated by the Model and Observed Values.**

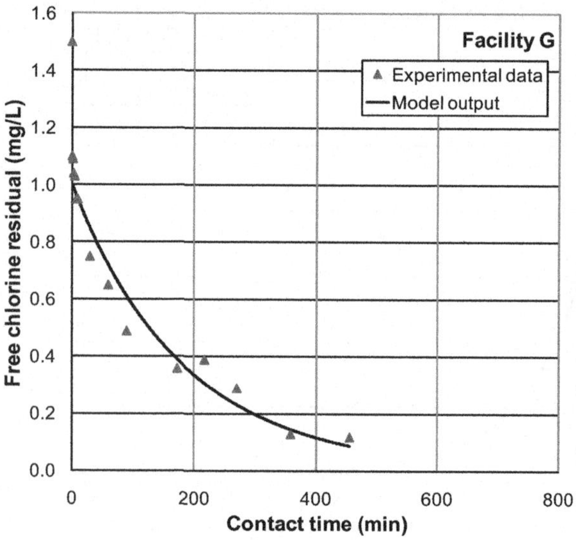


**Figure 4-1 continued. Comparison of Bulk Fluid Chlorine Concentrations Calculated by the Model and Observed Values.**

The model produced bulk fluid chlorine concentrations that were a good fit for the experimental data. In some of the tests there was more variability in the experimental data than in others. For example, the chlorine data collected for Facilities D, F, and G appears to have considerable variability. There also appears to be sampling, possibly the result of diluting samples, in some of the tests. However, the model appears to fit the shapes of the bulk fluid chlorine curves well. In addition to bulk fluid chlorine concentrations calculated by the model, concentrations of occluded total coliform were also calculated using the model (Figure 4-2). The model fit for occluded total coliform concentrations do not fit the observed values as closely as what was obtained for the bulk fluid chlorine concentrations. There appears to be more variability in the total coliform data than in the chlorine data (e.g., Facilities B and D). The model does not fit the shape of the data as closely either. Most of the total coliform concentrations appear to have a linear trend when plotted on a log-scale. However, the model produces total coliform concentrations that have a lag or shoulder phase followed by a steep linear section. Most of the total coliform lines are highly non-linear and have different sections with different local slopes. The model provides a fairly good fit for the initial observations, but this is because the initial concentration of occluded total coliform bacteria is one of the model inputs. The weights for the observations were altered to obtain a better fit for the final total coliform observations because the final values in a disinfection test are the most important.

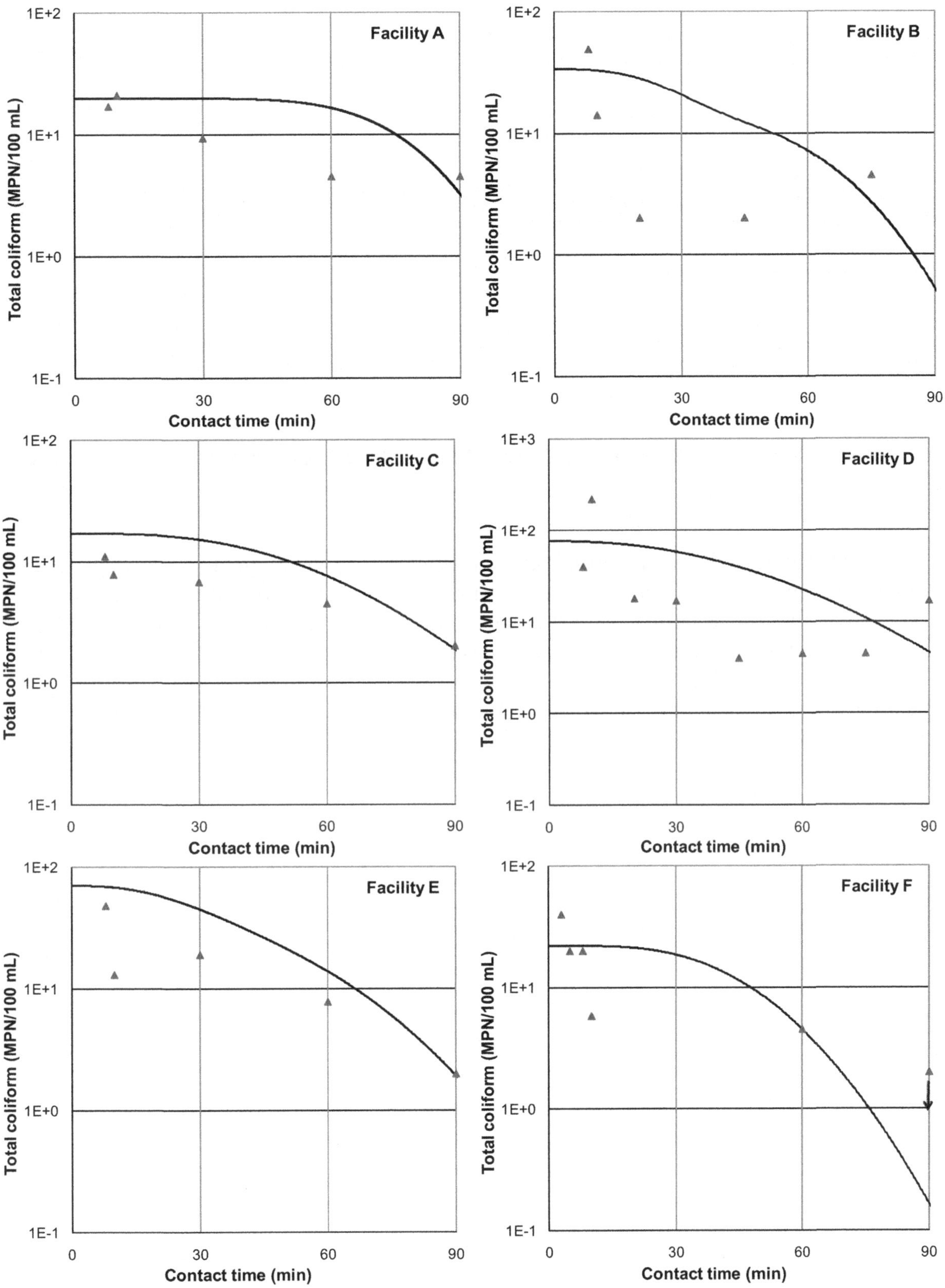


**Figure 4-2. Comparison of Occluded Total Coliform Concentrations Calculated by the Model and Observed Tailing Values.**

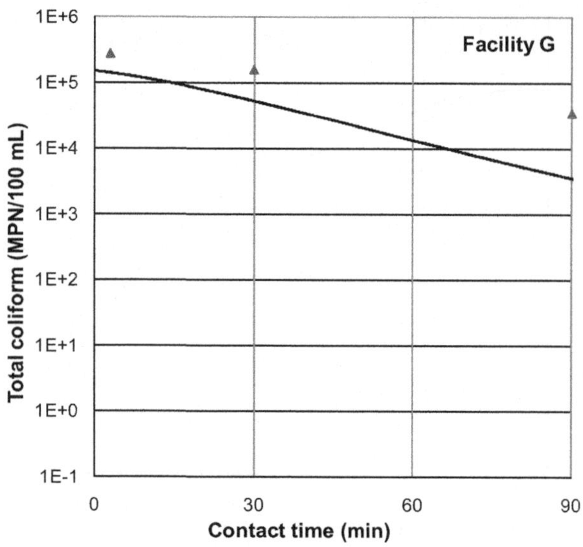


**Figure 4-2 continued. Comparison of Occluded Total Coliform Concentrations Calculated by the Model and Observed Tailing Values.**

The odd shapes of the total coliform curves generated by the model (Figure 4-2) are the result of the distribution of occluded total coliform bacteria across the particle size classes. To better understand the role of particle size and inactivation of occluded total coliform bacteria, data were generated by the model that were based on the assumption that there is at least one total coliform bacteria in every particle larger than 11 µm and that there is a total coliform bacteria at the center of each of these particles (Figure 4-3). In Figure 4-3 Particle 1 is the smallest particle in the model and Particle 4 is the largest particle.

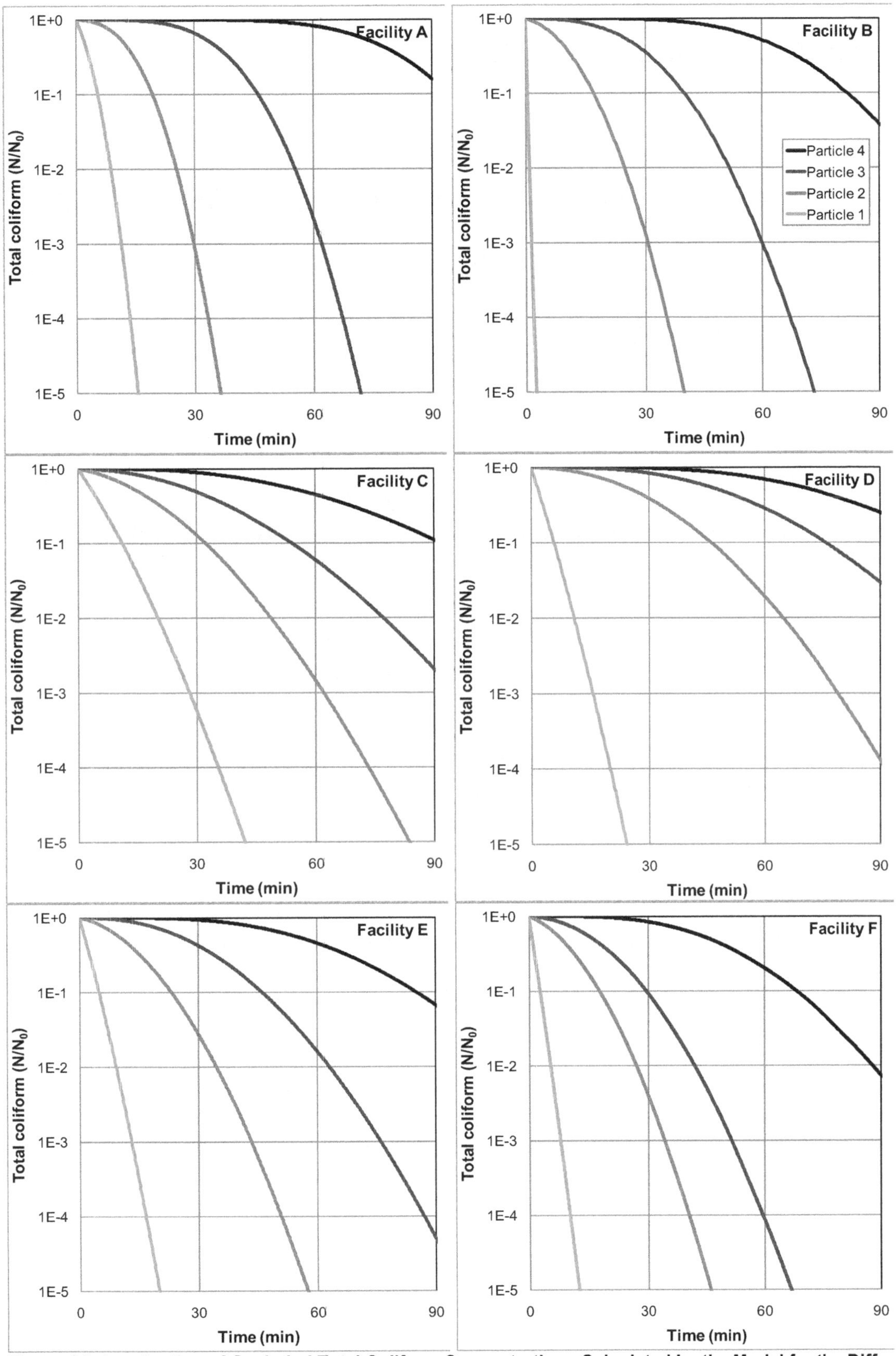


**Figure 4-3. Inactivation of Occluded Total Coliform Concentrations Calculated by the Model for the Different Size Particles, Assuming There is at Least One Occluded Total Coliform Bacterium in Each Particle Size Class Initially.**

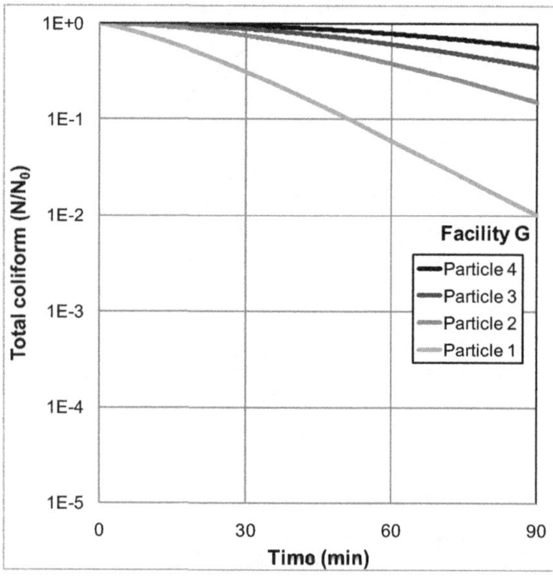


**Figure 4-3 continued. Log10 Inactivation of Occluded Total Coliform Concentrations Calculated by the Model for the Different Size Particles.**

In addition to calculating the bulk fluid chlorine concentration, the model calculates intra-particle chlorine concentrations (Figure 4-4). In Figure 4-4 the intra-particle chlorine profile are shown for all four particles sizes 10 and 90-minutes after the addition of chlorine. The profiles extend from the particle centers to the particle surface. As shown in Figure 4-4, the shape of the chlorine profile within particles is dependent on the size of the particle, the disinfection characteristics that differ for each facility, and the contact time. Typically, the chlorine concentration is higher at the particle surface than at the center of the particle, indicating that chlorine is diffusing into particles. In some cases, however, the chlorine profile is relatively flat, indicating a small concentration gradient inside of particles. For many of the larger particles, chlorine has not fully penetrated all regions of the particle interiors and the concentrations at the particle center are negligible. At a given time step, chlorine concentrations at the surfaces of different size particles is different, indicating differences in mass transfer at the particle surface. Where the profiles are relatively flat, as the concentrations at the particle surfaces for the different size particles are more similar.

The intra-particle chlorine profiles for Facility A have the steepest slope, indicating the fastest rate of mass-transfer. In contrast, the profiles for Facility G are nearly flat. For the model output for Facility A, the increase in intra-particle chlorine concentrations for each particle size appears dependent on the particle size. The smallest particle size had the largest increase in concentrations between 10 and 90-minutes of disinfection contact time. For Facilities B, D, E, and F the increases in concentration were less extreme.

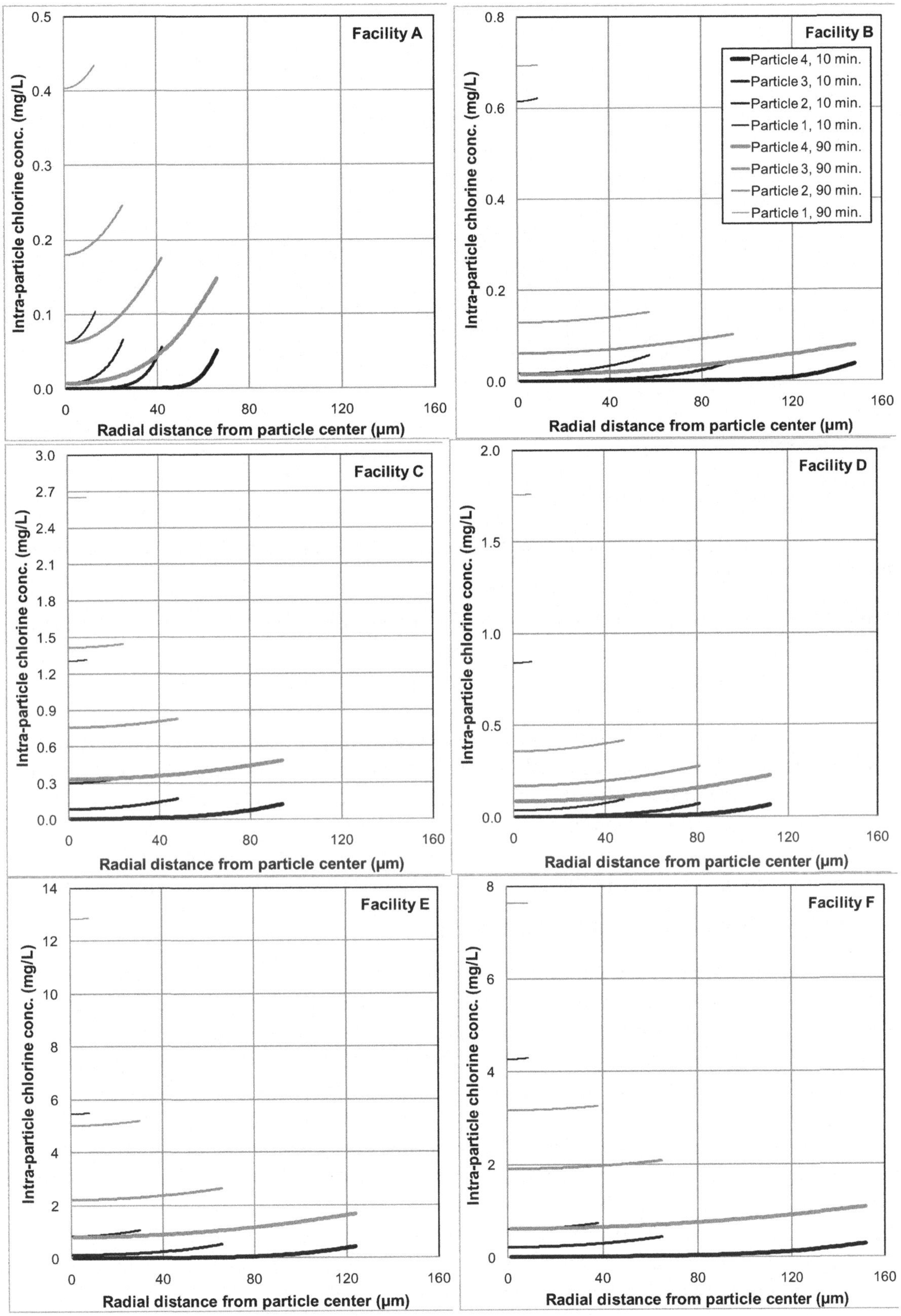


**Figure 4-4. Intra-Particle Chlorine Concentrations Calculated by the Model for the Center of Particles 10 and 90-minutes after Chlorine Addition.**

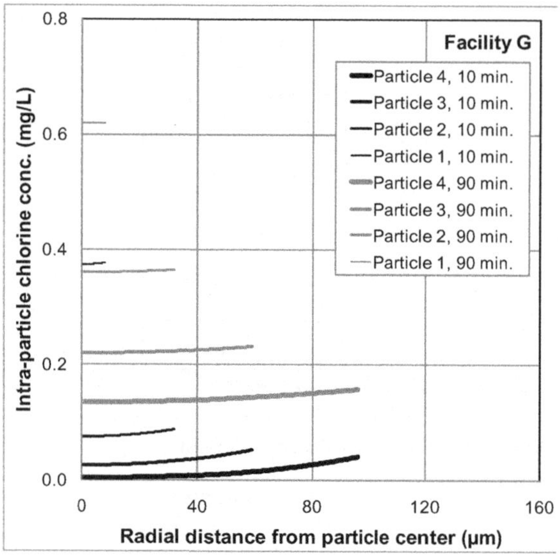


**Figure 4-4 continued. Intra-Particle Chlorine Concentrations Calculated by the Model for the Center of Particles 10 and 90-minutes after Chlorine Addition.**

An important model output in this study is the concentration of occluded total coliform bacteria over time as the microorganisms are exposed to increasing concentrations of chlorine. The relevant intra-particle chlorine concentrations are dependent on the initial quantity and distribution of occluded total coliform bacteria across the particle size classes. For example, in the case of Facility A there were only 20 occluded total coliform per 100 mL initially and all of these occluded microorganisms were assumed to inhabit the centers of the largest particles. The intra-particle chlorine concentrations for the smaller particles were not relevant in calibration of the model. In data sets where there were more occluded total coliform bacteria initially and where more of these microorganisms were assumed to inhabit the centers of smaller particles, the intra-particle chlorine concentrations in the different particle sizes becomes more important.

## 4.6 Sensitivity of Model Inputs and Parameters

Part of the functionality of the parameter estimating software (UCODE) is that it calculates sensitivities of the model output to small changes in the designated model parameters. By calculating sensitivities, the software determines the best choices for model parameters in successive model run iterations, making it possible for the output to converge on an optimal solution. Following model calibration, the parameter estimating software was run using the calibrated model for Facility A, Experiment A-1, to determine the sensitivities of several of the model inputs (Table 4-4). Most of these inputs were not previous included as parameters, but were temporarily added to determine the sensitivity of the model output to small changes in these inputs and model parameters. The scaled sensitivities were scaled by the parameter values. The purpose of this exercise was to determine the sensitivity of the model to assumptions made and data used for the Biot number, inactivation rate coefficient, reference particle size, and particle porosity. It should be noted that the sensitivities shown in Table 4-4 are based on the starting values for the inputs and parameters and that the sensitivities would change if the starting values were altered. For example, the intra-particle chlorine decay rate would be more influential if it were a higher number. The starting values selected were those used in the calibrated model.

**Table 4-4. Sensitivities of Model Inputs and Parameters for Facility A, Experiment A-1.**

| | Parameter starting value | Composite scaled sensitivities | Ratio of sensitivity to the maximum sensitivity |
|---|---|---|---|
| Biot number (*Bi*) | 0.1 | 6.8E-02 | 1.1E-02 |
| Inactivation rate coefficient ($k_i$) (L/mg*min) | 13.9 | 9.1E-02 | 1.4E-02 |
| Intra-particle diffusion coefficient ($D_a$) ($cm^2$/min) | 0.0006 | 6.5E-01 | 1.0E-01 |
| Reference particle radius (*R*) (μm) | 66 | 5.0E-02 | 7.9E-03 |
| Particle porosity ($\theta$) | 0.86 | 3.4E-02 | 5.4E-03 |
| Decay rate inside of particles (*k*) ($min^{-1}$) | 0.003466 | 0.0E+00 | 0.0E+00 |
| Ratio used to calculation $\eta$ ($k_f/k_m$) ($cm^{-1}$) | 100 | 2.6E-02 | 4.0E-03 |
| Partitioning coefficient ($K_d$) | 18670 | 4.0E-02 | 6.3E-02 |
| Decay rate in the bulk fluid (*k'*) ($min^{-1}$) | 0.003466 | 6.3E+00 | 1.0E+00 |

The results of the UCODE output shown in Table 4-4 indicate the importance of identifying appropriate inputs for the model during calibration. The model parameter for chlorine decay rate in the bulk fluid, *k'*, was the most influential parameter while there was no sensitivity for the chlorine decay rate inside of particles, *k*. The lack of influence for the decay rate inside of particles can be explained by the very high value for $K_d$, the sorption coefficient, which suggests that intra-particle chlorine decay is not influent for the relative values of *k* and $K_d$. The sensitivities for model inputs for Biot number, inactivation rate coefficient, and $K_d$ were similar. The relative sensitivities for the reference particle radius, particle porosity, and the ratio used to calculate $\eta$ were an order of magnitude lower than the relative sensitivities for the Biot number, inactivation rate coefficient, and $K_d$. The results shown in Table 4-4 demonstrate the interchangeable nature of the Biot number and $K_d$ for controlling intra-particle chlorine conditions and inactivation inside of particles. The results also demonstrate the importance of determining an appropriate value for the inactivation rate coefficient and that the particle porosity is not a very influential model input. However, the results of the sensitivity analysis indicate that the intra-particle diffusion coefficient is a very influential model input, which suggests that considerable care should be taken in selecting its value.

## 4.7 Model Simulation Exercise Using Data from Facility A

Three additional samples were collected from Facility A to complete a model simulation exercise. Experiment A-1 represents the work that was completed originally as part of the evaluation of the seven facilities and the other three data sets are referred to as Experiments A-2 through A-4. The model simulation exercise was conducted using two approaches. First, the calibrated model, which was calibrated using data from Experiment A-1, was used to run simulations with the PSD data observed and the initial chlorine concentrations used in Experiments A-2 through A-4. For each experiment the instantaneous chlorine demand was assumed to be the same for each of the disinfection tests. The inactivation coefficient measured in Experiment A-1 was used in all simulations. The purpose of the first simulation exercise was to determine how well the model fit the observed data when only the PSD and chlorine concentrations were known. The second simulation exercise consisted on calibrating the model separately for each experiment to determine the range of values obtained for the model parameters. As a comparison, additional simulations were run using the model that was calibrated separately for each experiment for the additional disinfection tests (data were only collected in the tailing region of the disinfection tests) to determine if there was an improvement in the model fit of the observed data. Data used in the model simulation exercise are shown in Table 4-5.

**Table 4-5. Model Simulation Exercise Data Used for Model Calibration.**

| | Experiment A-1 | Experiment A-2 | Experiment A-3 | Experiment A-4[1] | |
|---|---|---|---|---|---|
| Particle diameters (μm) | 26<br>50<br>84<br>132 | 18<br>52<br>90<br>124 | 18<br>46<br>90<br>204 | 16<br>56<br>80<br>100 | 14<br>62<br>80<br>118 |
| Particle concentrations ($\omega_i$) (#/100 mL) | 2941<br>1806<br>114<br>20 | 3753<br>293<br>50<br>3 | 6475<br>537<br>82<br>5 | 6948<br>167<br>16<br>5 | 1724<br>13<br>6<br>6 |
| Total experiment time in the disinfection test used to calibrate the model (min) | 489 | 418 | 343 | 402 | |
| Particle porosity ($\theta$) | 0.86 | 0.86 | 0.00 | 0.80 | |
| Initial concentration of occluded total coliform ($N_0$) (MPN/100 mL) | 0<br>0<br>0<br>20 | 0<br>141<br>50<br>3 | 0<br>0<br>81<br>5 | 124<br>167<br>16<br>5 | 0<br>0<br>0<br>6 |
| Inactivation rate coefficient ($k_i$) (L/mg*min) | 13.9 | 23.1 | 20 | 35.1 | |
| Initial chlorine concentration[2] ($C_{bulk,t=0}$) | 4.9 | 4 | 2.8 | 1.7 | |

[1]One of the disinfection tests conducted during Experiment A-4 was filtered prior to the test and had a different PSD. The PSD for the unfiltered sample is shown in the left column and the PSD for the filtered sample is shown in the right column.
[2]Initial chlorine=chlorine added - instantaneous chlorine demand. Only the initial chlorine concentration for the full set of data necessary to calibrate the model is shown. Additional disinfection tests were performed.

Similar to what was done for the seven facilities, the PSD data were divided into four particle size classes and the initial concentrations of occluded total coliform bacteria were identified by examining the data collected in the disinfection tests. A best-fit line was drawn through the total coliform data in the tailing region of the test and the line was extended to the y-axis that represents the zero dose or initial conditions. The y-intercept of the best-fit line was assumed to be the initial concentration of occluded total coliform bacteria. The initial concentrations of occluded total coliform bacteria were then distributed over the four particle size classes. The inactivation rate coefficients for dispersed total coliform bacteria were measured separately on each of the days when data were collected, and the results vary considerably. The initial chlorine concentrations used to calibrate the model were determined by viewing the chlorine data collected during the disinfection tests, and represent the amount of chlorine added to the reactors less the instantaneous chlorine demand. During Experiments A-2 to A-4 more than one disinfection test was performed; however, on each day only one disinfection test was conducted where enough data were collected to calibrate the model. During the additional disinfection tests sufficient data were only collected to compare the data with model simulation results. For example, data were only collected during the tailing portion of the disinfection tests (more than 10 minutes after the addition of chlorine).

In the first part of the simulation exercise the model, as calibrated using the data from Experiment A-1, was used to run simulations using the PSD, initial total coliform, and initial chlorine concentrations from the subsequent experiments (Figure 4-5). For Experiment A-2, the model estimates of total coliform bacteria were higher than what was actually observed during the experiment. However, in the case of Experiment A-4, the model estimates for total coliform bacteria were lower than what was observed. The model provided a relatively good fit for the data collected in Experiment A-3.

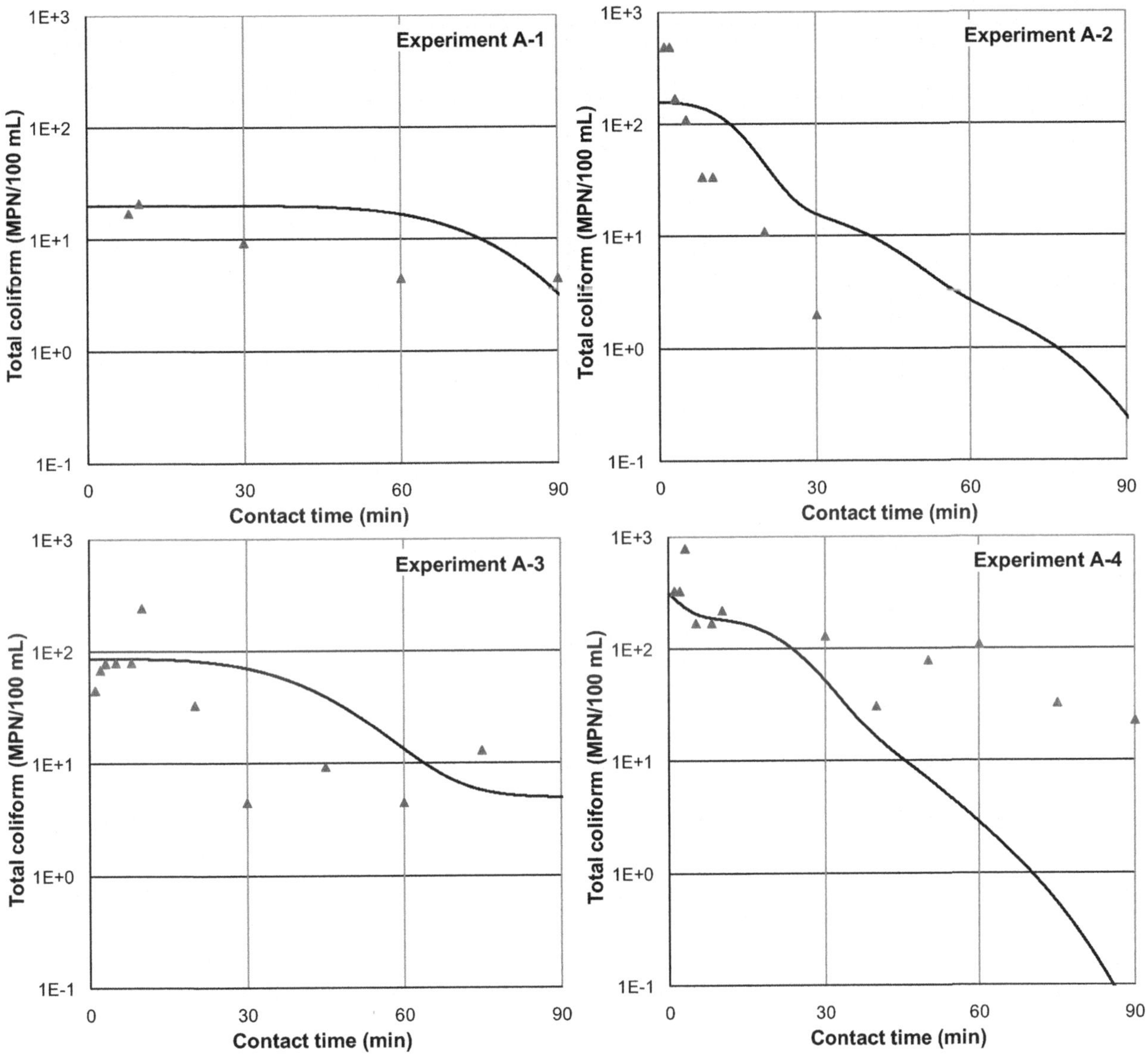


**Figure 4-5. Model Output for Occluded Total Coliform Bacteria Using the Data from the Model Simulation Exercise Experiments and the Model Parameters from Calibration Using Experiment A-1 Data (k=k'=0.003466 min$^{-1}$ and Kd=18,670).**

Model inputs and parameters for the calibrated models completed as part of model simulation exercise are indicated in Table 4-6. The information in Table 4-6 was collected in the second part of the simulation exercise where the model was calibrated separately for each day that experiments were conducted for Facility A. The intra-particle diffusion coefficient used in all experiments was 0.0006 cm2/min. Similar to what was done for calibration of the model for the seven facilities, 101 nodes were used and the dimensionless time step was 0.01. In addition, a value of 0.1 was assigned to all of the Biot number parameters and the ratio *kr*/*km* was assumed to be 100 $cm^{-1}$.

**Table 4-6. Model Simulation Exercise Model Inputs and Parameters.**

| Model parameter | Experiment A-1 | Experiment A-2 | Experiment A-3 | Experiment A-4 |
|---|---|---|---|---|
| Reference particle (*R*) (µm) | 132 | 124 | 204 | 100 |
| Dimensionless radius ($R_i$) (µm) | 0.20<br>0.38<br>0.64<br>1.0 | 0.15<br>0.42<br>0.73<br>1.0 | 0.09<br>0.23<br>0.44<br>1.0 | 0.16<br>0.56<br>0.80<br>1.0 |
| Number of dimensionless time steps | 674,000 | 653,000 | 198,000 | 965,000 |
| Biot number (*Bi*) | 0.1 | 0.1 | 0.1 | 0.1 |
| Ratio of the desorption rate to the mass-transfer coefficient across the particle boundary ($k_r/k_m$) (1/cm) | 100 | 100 | 100 | 100 |
| Desorption to mass transfer at the particle surface ratio ($\eta$) | 0.57<br>0.36<br>0.22<br>0.11 | 0.54<br>0.39<br>0.23<br>0.08 | 0.89<br>0.39<br>0.20<br>0.08 | 0.44<br>0.35<br>0.24<br>0.07 |
| Ratio of the sorption rate to the desorption rate ($K_d$) | 18,670 | 7759 | 8831 | 64,260 |
| Decay rate inside and outside of particles ($k=k'$) (1/min) | 0.0035 | 0.0029 | 0.0049 | 0.011 |
| Intra-particle Thiele modulus ($\varphi$) | 0.005<br>0.003<br>0.002<br>0.001 | 0.005<br>0.003<br>0.002<br>0.001 | 0.010<br>0.004<br>0.002<br>0.001 | 0.007<br>0.006<br>0.004<br>0.001 |
| Thiele modulus for the external bulk fluid ($\varphi'$) | 0.005 | 0.005 | 0.010 | 0.007 |

As shown in Table 4-6 the value of the model parameter $K_d$ varied widely for the four experiments. The values for $K_d$ in Experiments A-1 and A-4 were high, 18,670 and 64,260, respectively. The values for $K_d$ in Experiments A-2 and A-3 were much lower and were fairly similar, 7759 and 8831, respectively. Values for $k=k'$ varied from 0.011 to 0.0049 $min^{-1}$. The calculated values for the Thiele moduli were fairly similar, and the intra-particle values varied according to the particle size. Model output for the bulk chlorine concentration and the observed values are shown in Figure 4-6.

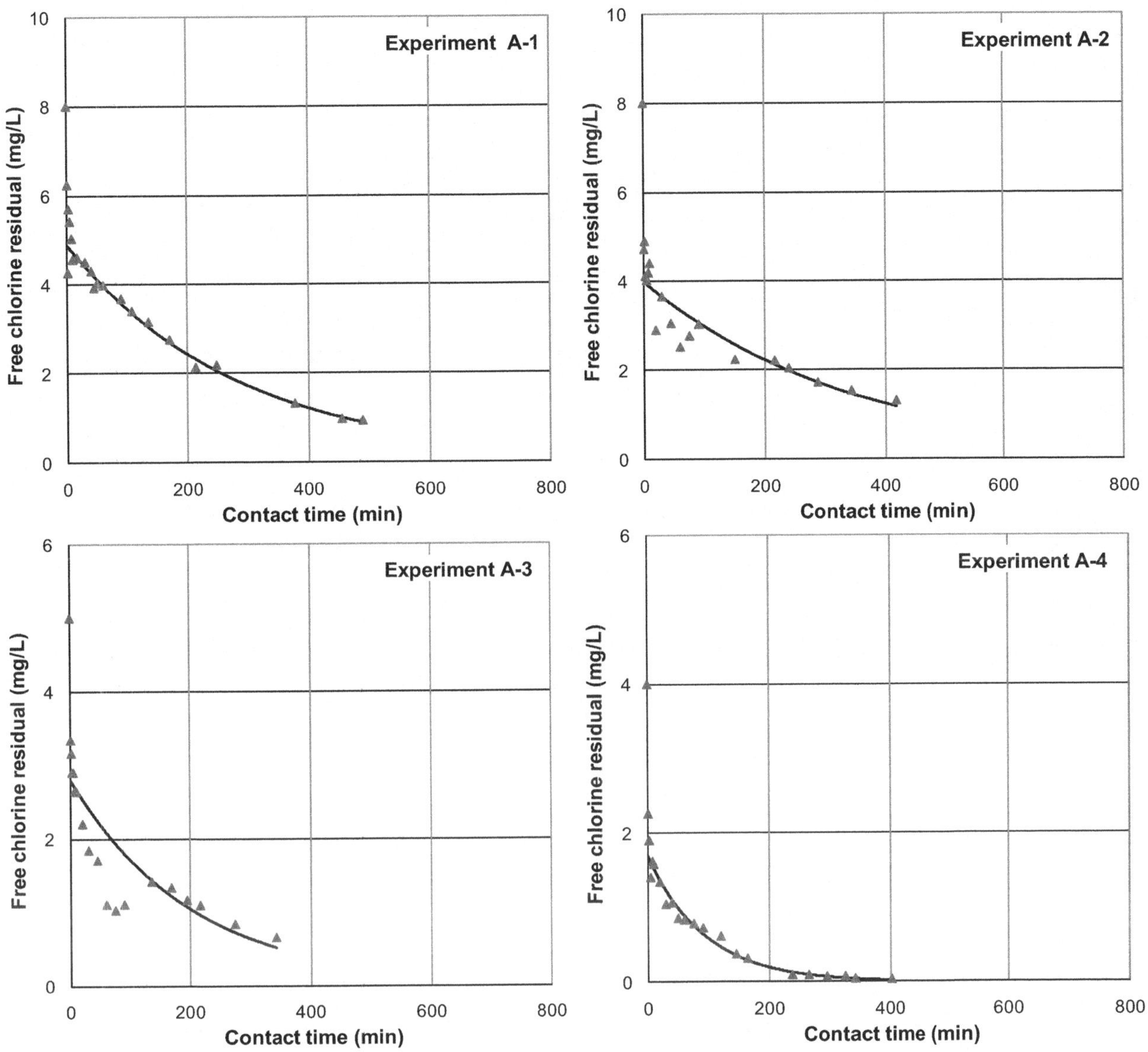


**Figure 4-6. Model Output for Bulk Fluid Chlorine Concentration Obtained Calibrating the Model Separately for Each Day That Data Were Collected for the Model Simulation Exercise.**

As was the case when data from the seven facilities were used to calibrate the model, the model provided a good fit for the observed bulk fluid chlorine data. The model output for occluded total coliform bacteria is shown in Figure 4-7 along with the observed total coliform concentrations in the tailing regions of the disinfection tests. When the data from the individual experiments was used to calibrate the model, good fits were obtained for the experimental data.

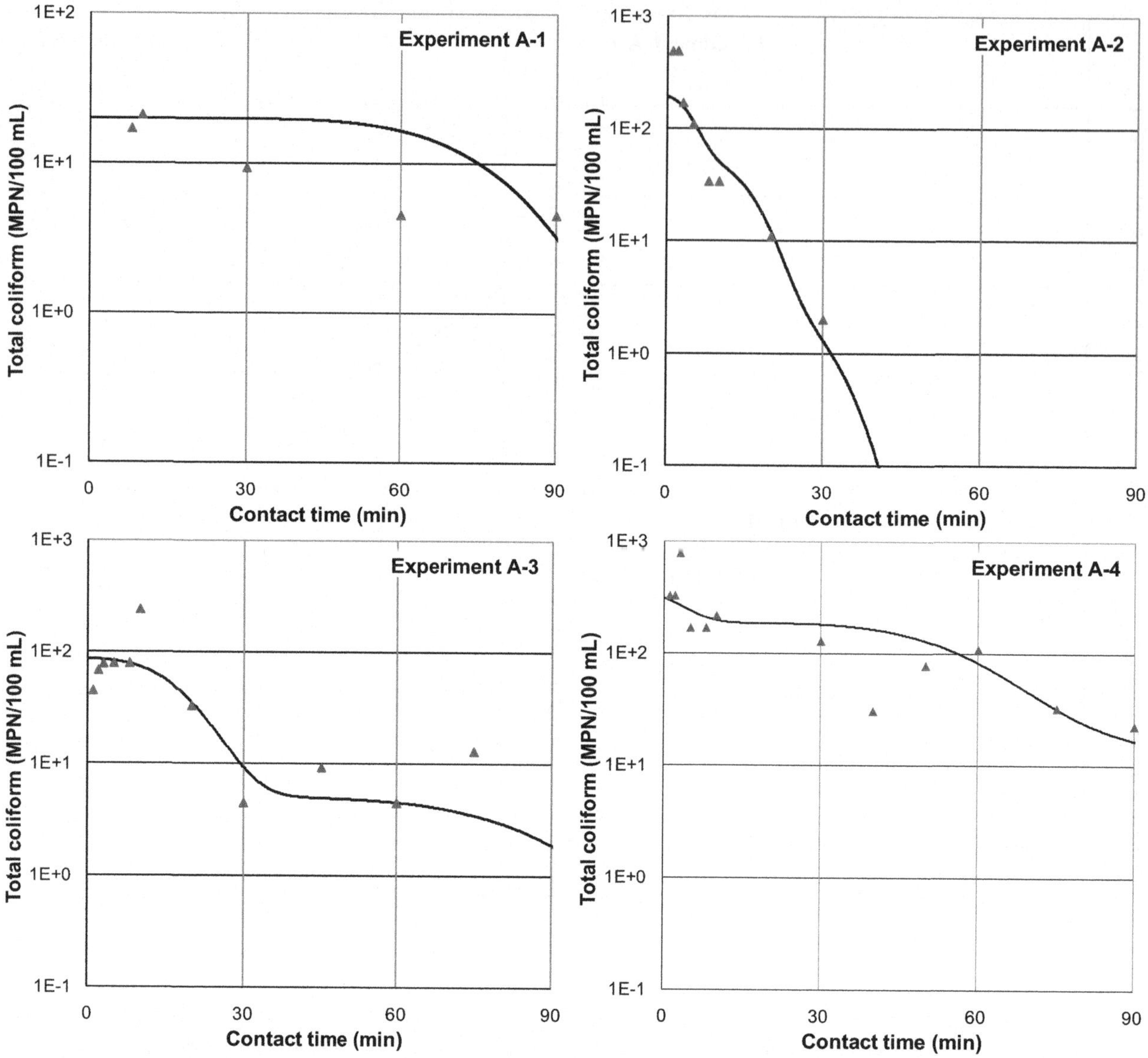


**Figure 4-7. Model Output for Occluded Total Coliform Concentration Obtained Calibrating the Model Separately for Each Day That Data Were Collected for the Model Simulation Exercise.**

After calibrating the model separately for each experiment, the data from the additional disinfection tests were plotted along with model simulation output that was obtained using the calibrated model with different initial chlorine concentrations (Figure 4-8). In the additional disinfection tests, a subset of the sample was used and the tests were conducted with different initial concentrations of chlorine. In one of the tests conducted during Experiment A-4 the subset of the sample was filtered through a 20 m net nylon filter in a sterile filter apparatus prior to the disinfection test. The PSD of the filtered sample was measured and this data was used in the model simulation in lieu of the PSD data shown in Table 4-5.

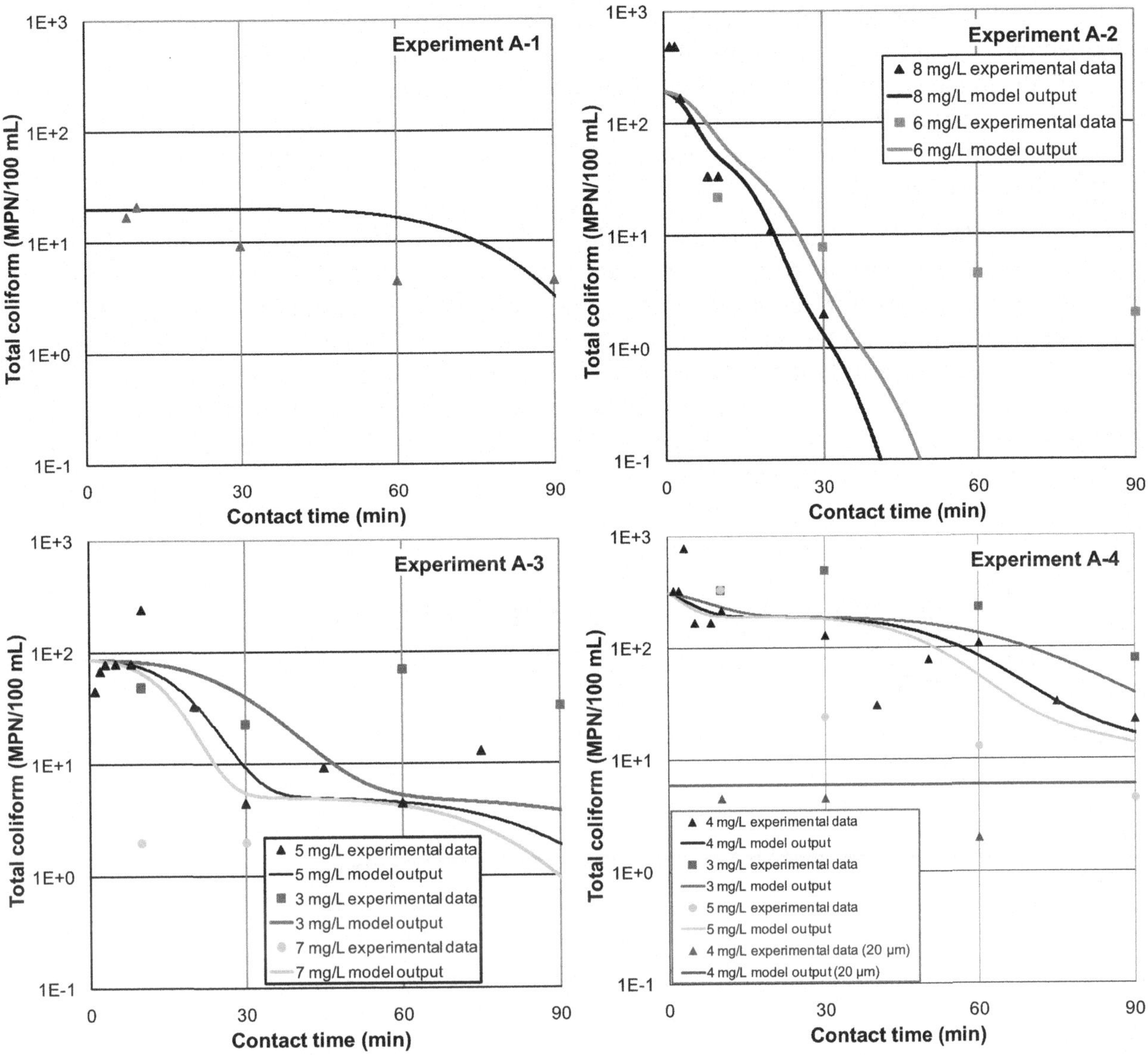


**Figure 4-8. Model Output for Occluded Total Coliform Concentration Obtained Calibrating the Model Separately for Each Day That Data Was Collected for the Model Simulation Exercise, Results Shown from Multiple Disinfection Tests Conducted Using Subsets of the Same Sample.**

In Experiment A-2, data from the disinfection test conducted using an initial chlorine concentration of 6 mg/L exhibited tailing that was not present in the data collected from the test with an initial chlorine concentration of 8 mg/L. The model did not provide a good estimate of the tailing in the test with 6 mg/L. In the test conducted using 8 mg/L data were not available for calibration for contact times greater than 30 minutes; all of the total coliform data collected after 30 minutes were below the detection limit of 2 MPN/100 mL. In Experiment A-3 the model simulation output for occluded total coliform bacteria were not appreciably different although the data collected in the three disinfection tests exhibited different tailing that corresponded to the initial concentrations of chlorine used in the respective tests. The result for Experiment A-4 was similar to what was observed for Experiment A-3. The model simulation results for varying initial chlorine concentrations were similar despite considerable differences in the degree of tailing observed in the disinfection tests. For the filtered sample used in Experiment A-4, the model simulation indicated very little inactivation whereas a modest amount of inactivation was observed during the disinfection test. Note that the initial concentration of occluded total coliform bacteria in the disinfection test using the filtered sample was different than in the other tests because filtering the sample resulted in the removal of a portion of the occluded total coliform bacteria.

The wide range of model parameters identified in separately calibrating the model for the data sets collected as part of the model simulation exercise may appear excessive; however, it is necessary to examine the effect of the model parameters on the model output. An example is shown in Figure 4-9 using the data and calibrated model from Experiment A-1 with different values of $K_d$. As shown in Figure 4-9 the curve for occluded total coliform bacteria is relatively flat when $K_d$ is higher than approximately 20,000. The differences in the model output become less significant as the value of $K_d$ is increased. For values around 20,000, the final occluded total coliform concentration is fairly stable.

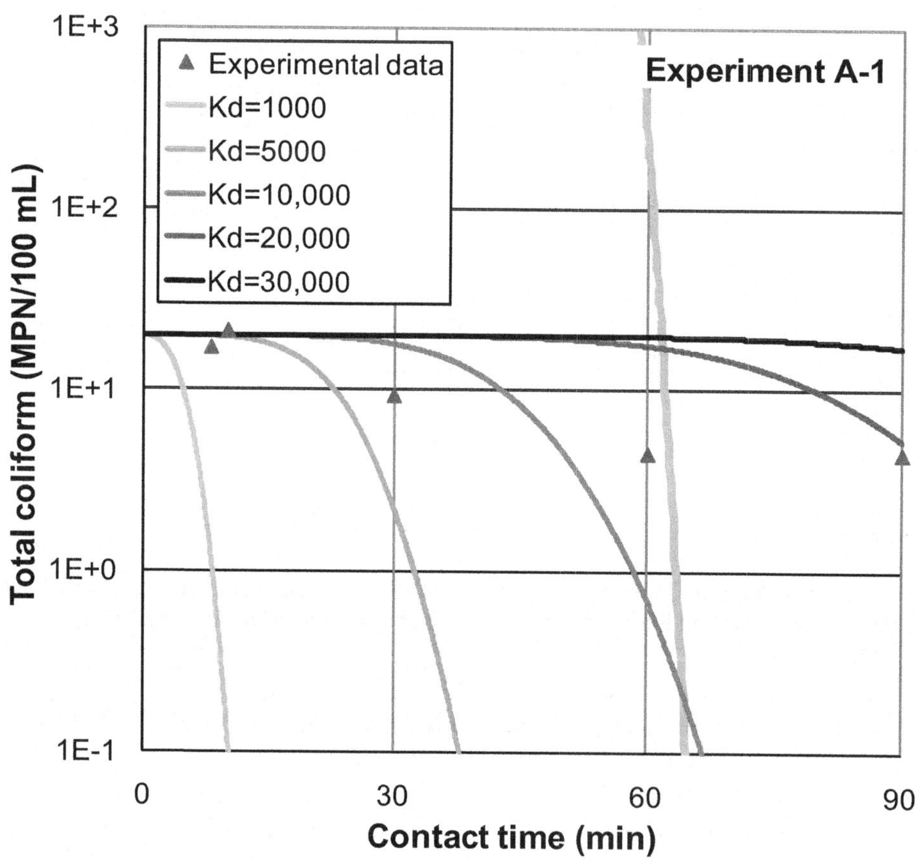


**Figure 4-9. Calibrated model for Facility A, Experiment A-1, using different values for $K_d$.**

As $K_d$ is increased, intra-particle sorption increases and the concentration gradient at the particle surface increases as well. The lower gradient induces additional mass transfer into particles and the bulk fluid decay rate does not need to be as high because chlorine is leaving the bulk fluid via mass transfer and not via decay. In summary, the model parameters that control decay and sorption work collectively to define the chlorine conditions inside and outside of particles. The Biot number is also important because it controls mass transfer at the particle surface, although in this study it was held constant.

## 4.8 Discussion and Conclusions

A radial pore diffusion model was used to gather information regarding intra-particle chlorine concentrations and the inactivation of microorganisms occluded within particles. The model considers an aqueous constituent, chlorine in this case, that diffuses from a well-mixed bulk fluid into spherical particles. Although the model was used in this study for inactivation of total coliform bacteria with free and combined chlorine, it is possible to use the model for inactivation of other microorganisms using any chemical oxidant. First-order decay kinetics were assumed for chlorine in the bulk fluid and inside of particles; first-order sorption kinetics inside of particles were also assumed. The model output consisted of intra-particle chlorine concentrations that were spatially distributed and changed following the introduction of chlorine. Although useful, the model has inherent limitations that are a result of the method in which the model was formulated and the assumptions that went into using the model. In the model four particle sizes were used and the particles were assumed perfectly spherical, whereas in the wastewater samples each particle is unique and there is a distribution of particle sizes. The model does not accommodate different chemical species and characteristics; a single form of chlorine was considered in each application of the model. To model chlorine reactions with non-soluble organic matter within particles, a linear sorption model was used and use of alternative models and/or alternative sorption models was not pursued.

The following laboratory data were used to run the model: PSD (used to determine $R_i$ and $\omega_i$ and the particle volumetric void ratio($\varepsilon_i$)), initial chlorine concentration in the bulk fluid ($c_{bulk,t=0}$), particle porosity, diffusion coefficient, inactivation rate coefficient for dispersed total coliform bacteria, and the initial concentration of occluded total coliform bacteria. Chlorine and total coliform observations recorded during the disinfection tests were used to calibrate the model. Parameter estimating software (UCODE) was used to calibrate the model, which functioned using a regression analysis and minimizing the sum of the squared residuals. Model parameters were selected that resulted in an optimal model solution, making it possible to compare model parameters selected for different facilities. Values for the diffusion coefficient ($D$) and Biot number were selected based on published values. The parameters $k=k'$ and $K_d$ were changed during calibration. The Biot number was set at 0.1, based on the lower end of the likely range of values. The decay rates ($k=k'$) were used in the model to calculate the Thiele modulus for the intra-particle aqueous phase ($\varphi$) and the Thiele modulus for the external bulk fluid ($\varphi'$). The ratio $k_r/k_m$ was set at 100 $cm^{-1}$; this ratio is used to calculate the desorption to mass transfer at the particle surface ratio ($\eta$). The values for $\varepsilon$, $\varphi$, $\varphi'$, $\xi$, and $\eta$ are dependent on the radius size of the particles and separate values were calculated for each of the four particle size classes.

An attempt to calculate $k_m$ based on literature values was made using Harriott (1962), but the calculated Biot numbers were too high and the mass transfer of chlorine from the bulk fluid into the particles was too rapid. It is likely that the experimental conditions from Harriott (1962) were not representative of the conditions in this study. Alternative methods for determining the Biot number are available, and additional work could be conducted to identify appropriate values. It would be preferable to determine appropriate values for $k_m$ that are based on the individual particle sizes. Identifying values for $k_m$ would allow the Biot number to be calculated independently for each particle size and it would allow $\eta$ to include $k_m$ in its calculation.

Identifying values of $k_m$ would allow $k_r$ to be a model parameter instead of $k_r/k_m$, which was assumed constant in this study. The parameter $k_r$ is based on sorption and it is not expected that it would vary for different particle sizes. It is possible that $k_r$ could be used as a fitting parameter as long as $k_r$ and $K_d$ were not highly correlated.

During model calibration assumptions were made regarding the Biot number and the ratio $k_r/k_m$, which affected the selection of model parameters. Had different assumption been made (e.g., higher Biot number) the selected model parameters would have been different. In this study, the calibration procedure and assumptions were kept consistent for all of the facilities so that the results could be compared. In general, $k'$ affects the bulk fluid chlorine concentration and $K_d$ impacts the intra-particle conditions (total coliform inactivation) although both parameters affect mass-transfer in the system. The parameter $k$ impacts the intra-particle conditions although $K_d$ was much more influential in this study because $k$ was constrained by $k'$, which was basically determined by the bulk fluid chlorine observations. To a certain extent $K_d$ and $\eta$ are interchangeable because both parameters impact the rate or sorption, and this is especially true where the sorbed concentrations in the particles are low. Differences between the two parameters might be more apparent for longer contact times. For the conditions in this study both $K_d$ and $\eta$ could not be used as fitting parameters because the two parameters were too highly correlated. For similar reasons, it was not possible to include the Biot number as a fitting parameter.

Model calibration is dependent on the chlorine and total coliform data that are collected in the disinfection tests, and sufficient disinfection test data are necessary. A well-defined tailing region in the disinfection test data is required, where total coliform concentrations were above the detection limits. Selection of an appropriate initial concentration of chlorine is needed to develop an inactivation curve with a tailing region. One of the major assumptions in this study was that tailing in the disinfection tests was the result of total coliform occlusion in particles. Other important assumptions made in this study were that the occluded total coliform bacteria occupy the center of particles and that they are occluded in larger particles prior to smaller particles. Both assumptions are conservative and assure that all regions of particles are sufficiently saturated with chlorine such that all occluded total coliform are inactivated.

In the disinfection test it would be desirable to have the tailing residual of total coliform as high as possible, which would ensure that the occluded total coliform occupy as many of the particle sizes as possible. The calibration would then be based on the intra-particle chlorine concentrations of all or most of the different size particles. As shown in Table 4-2, the initial concentrations of occluded total coliform bacteria are low and in most cases the occluded total coliform bacteria occupy only the largest particles. Where only the largest particles contain total coliform bacteria, the calibration is based on selecting model parameters that result in the appropriate intra-particle conditions for the largest particles. If, in subsequent model simulations, the intra-particle chlorine concentrations of the small particles are of interest, the model may not be adequately calibrated to provide good results for inactivation in smaller particles.

Using parameter estimating software offered several advantages over manually calibrating the model. Calibration was completed in a systematic manner that was consistent for all of the data sets. It was also possible to determine the sensitivities of the model output to small changes in the model parameters. In this way it was possible to identify the most strongly influential model parameters and to identify which model inputs are important. For example, a sensitivity analysis was used to determine that highly accurate values for particle porosity were not essential, but that care should be taken in selecting the most appropriate values for the diffusion coefficient. The parameter estimating software provides information about model parameter correlation coefficients and provides residual analysis. The contributions of individual observations to the objective function were reported, which helped in identifying observations for which the model was not providing a good fit.

There were some trends between the selected model parameters and the operating conditions at the facilities (e.g. MCRT). The partitioning coefficient ($K_d$) was much higher for Facility A than for the other facilities, while all of the other values were fairly consistent. One possible reason that the $K_d$ parameter was so much higher for Facility A was that the values for $\eta$ were lower than for the other facilities. The value of $k=k'$ was higher for Facility B than for the other facilities, which was expected since free chlorine is the disinfectant and it decays faster than combined chlorine that is relatively stable in environmental waters. It is possible that the value of $k=k'$ for Facility A would be higher if the value of $K_d$ was lower. Where sorption is more prevalent, the decay rates decrease because the decreasing concentration gradient across the particle surface (caused by increasing sorption) causes more chlorine to transfer into the particles. Where there is significant mass transfer into the particles, the decay rates are lower because decreases in the bulk fluid chlorine occur and chlorine decay in the bulk fluid is not as important. In this study chlorine decay rates were lower for facilities that have partially-nitrified and non-nitrified effluents, as was expected because combined chlorine is more stable than free chlorine.

Since the inactivation coefficients for the nitrified effluents are much higher than for the other facilities, it would be expected that intra-particle inactivation would be rapid. However, tailing is observed for the nitrified effluents in addition to tailing observed for the other facilities. In the case of Facility A the tailing appears to be related to reactivity of the intra-particle chlorine with the particle organic matter. In the case of Facility B, the prevalence of tailing appears to be related to reactivity of the chlorine in solution, as evident by the higher decay rate. The results for Facilities A and B indicate that free chlorine inside of particles is less effective for inactivating occluded microorganisms because of reactivity with organic matter and high decay rates.

The calibrated model provided relatively good estimates of the bulk fluid chlorine concentrations although the estimates for occluded total coliform bacteria were not as good. One of the issues with the model estimates for occluded total coliform was that the assumption that occluded total coliform are present in the centers of particles results in curves that have a lag or shoulder region at the beginning of the disinfection tests. The observations from the disinfection tests, however, indicate a linear relationship between occluded total coliform bacteria and time when the data are plotted on a log scale. Changing the assumption of total coliform bacteria locations within particles might result in curves that are more linear and more closely aligned with the observational data. Possible assumptions include assuming that total coliform bacteria occupy a range of particle sizes and not just the largest particles. Another possible assumption is that total coliform bacteria are located randomly throughout particles and not just at the particle centers. An additional scenario is that total coliform bacteria are located halfway between the particle surface and particle center.

Insufficient data were available to adequately calibrate the data collected for Facility G. Specifically, the initial chlorine concentration used in the disinfection test was inadequate to produce an inactivation curve with a tailing region that consisted of total coliform concentrations that were lower than the particle concentrations, suggesting that not all of the dispersed total coliform bacteria were inactivated in the disinfection test. The data from Facility G were used in the calibration and simulation procedures to determine approximate values. It was assumed that all particles contained at least one total coliform bacterium. A value of $K_d$=1500 was assigned to the Facility G data, which was based the range of values observed for the other facilities. The model parameter *k=k'* was selected using the parameter estimating software.

The log removal information for different particle sizes (Figure 4-3) is particularly insightful because it reveals how long it will take to inactivate total coliform occluded in various sized particles. For example, there is often a desired treatment objective (e.g. 5-log removal) and appropriate disinfection conditions must be selected to achieve this objective. For all of the facilities, there was less than approximately 2-log inactivation of total coliform bacteria occluded in the largest particles. Most of the smaller particles had rapid inactivation of the occluded total coliform bacteria and diffusion limitations were not apparent.

The chlorine concentration profiles at different time intervals reveal information about the diffusion of chlorine into particles and the resulting potential for inactivation of potentially present microorganisms. Diffusion of chlorine into particles is a dynamic process because diffusion is dependent on the concentration of chlorine in the bulk solution, which is continuously changing over the course of a disinfection test. In some cases the intra-particle chlorine concentrations dramatically increased over the course of the disinfection period. However, the larger particles still had regions without substantial concentrations of chlorine following 90-minutes of contact time. Had the disinfection period been longer, it is possible that chlorine would have started to diffuse out of particles as a result of the decreasing chlorine concentration in the bulk solution.

As a result of the boundary layers surrounding the particle surface, there is a discontinuity in chlorine concentration from the bulk fluid to the inside of particles, and this is reflected when the bulk fluid chlorine concentrations (Figure 4-1) are compared with the intra-particle chlorine concentrations (Figure 4-4) at the appropriate time step. Diffusion of chlorine into particles does not occur unless there is a concentration gradient across the particle boundary that favors diffusion into the particles. The diffusing chlorine must also overcome the energy potential associated with the boundary layer. The intra-particle chlorine concentrations can become equivalent to the bulk fluid concentration as bulk fluid concentrations decreases, but it is not an equilibrium situation. The bulk fluid chlorine concentration will decrease quicker than the intra-particle chlorine concentrations that change slowly due to diffusion characteristics in particles.

The model simulation exercise to run simulations using the model calibrated with the original data set for Facility yielded estimates of occluded total coliform bacteria that approximated the observed values and did not consistently overestimate or underestimate the occluded total coliform values. Subsequent calibration of the model simulation exercise data did not result in model parameters with a clear trend. Of the four sets of data, two yielded estimates of $K_d$ that were high (18,670 and 64,260) and the other two yielded estimates of $K_d$ that were much lower (7759 and 8831) although they were higher than the values identified for the other facilities (the range of values was 1429 to 2656). The values for *k=k'* were fairly consistent although the value for Experiment A-4 was slightly higher. The wide range of values for $K_d$ might suggest that the disinfection characteristics for a single facility are not consistent. However, the sensitivity analysis demonstrated that widely varying values of $K_d$ in the range of 15,000 to 30,000 for Experiment A-1 did not have a strong effect on the predicted effluent occluded total coliform concentrations. In general, high values of $K_d$ indicate that there is a tendency for chlorine to sorb to the particle interior surfaces.

When the separate data sets from Facility A were calibrated separately, good fits were obtained for the bulk fluid chlorine and total coliform observations. When the separately calibrated models were used to run simulations using PSD and initial chlorine data from additional disinfection tests, the results for occluded total coliform were similar. In the disinfection tests, however, small changes in the initial chlorine concentration yielded very different tailing concentrations of total coliform bacteria. It is possible that better fits would be obtained if different assumptions were made regarding the occlusion of total coliform in particles and the locations of these occluded total coliforms.

Several issues became apparent during the modeling process. The model provided a much better fit for the bulk fluid chlorine data than for the occluded total coliform data. The model output for the occluded total coliform bacteria more closely fit the initial data points than the final points because the initial total coliform concentration was a model input. There are several things that could be done to improve the fit of the total coliform data. One is to collect total coliform data for a period of time that is longer than the time period of interest. For example, if a 90-minute contact time is of interest, it would be preferable to collect two hours of data. It is possible that obtaining values for km and calculating the Biot number would result in better model estimates and model parameters exhibiting trends with the facility operating conditions. Exploring different assumptions for occlusion in particles and locations of occluded total coliform might result in better model estimates for inactivation of total coliform occluded in particles. Obtaining sufficient total coliform data in the tailing region of a disinfection test is crucial for model calibration. In addition, the initial concentration of occluded total coliform bacteria is very important and it can be identified from viewing the disinfection test data.

Intra-particle chlorine concentrations calculated by the model offer an opportunity to consider inactivation of microorganisms other than total coliform bacteria that may potentially be present in particles such as pathogens (Chapter 6.0). Additionally, using model simulations the effects of varying particle and initial chlorine concentration conditions can be evaluated (Chapter 5.0).

## CHAPTER 5.0

# Optimization Protocols for Filtration and Disinfection Systems

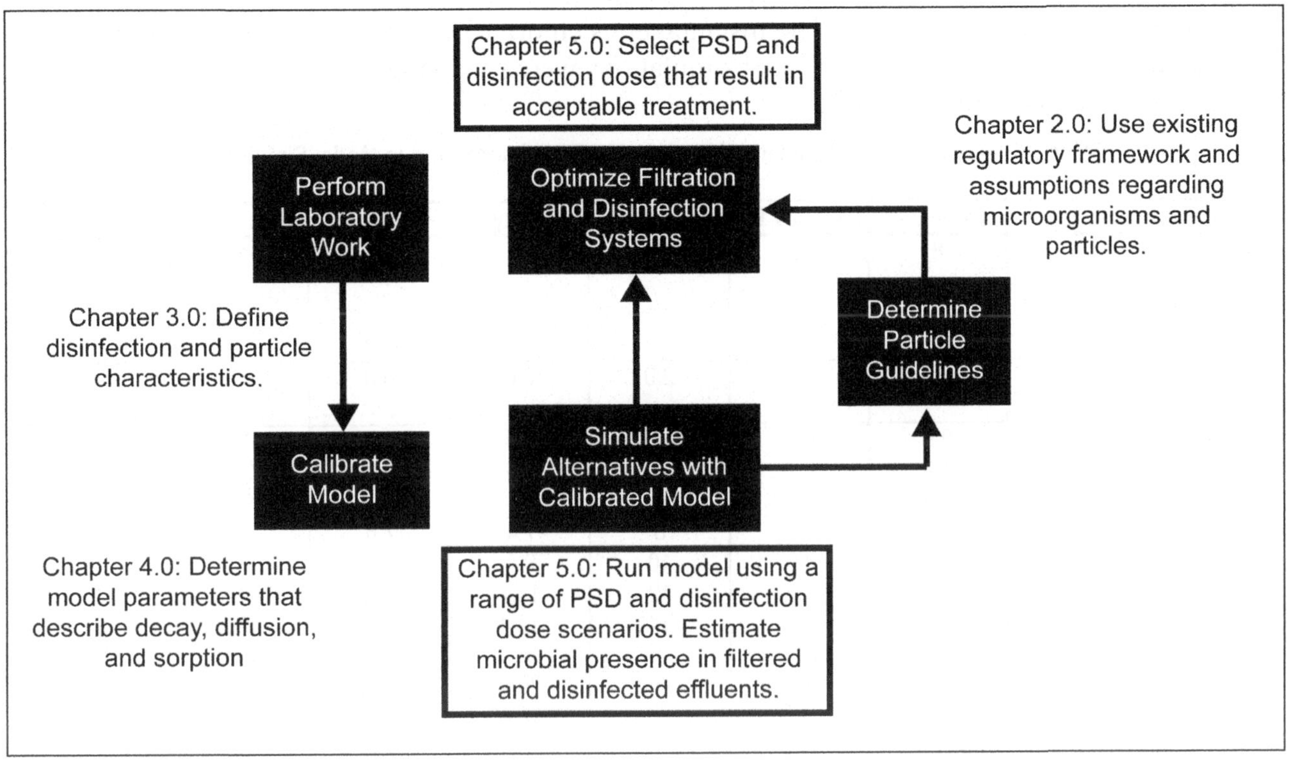


## 5.1 Introduction

For each facility, particle guidelines were developed based on the work presented in Chapter 2.0 and on the measured frequency of association (F) values that were presented in Chapter 3.0. The particle guidelines were compared with the measured particle counts also presented in Chapter 3.0. Particles counts were higher than the particle guidelines, and the model was used to determine how the filtration and disinfection systems could operate in tandem to collectively reduce the presence of target microorganisms in the effluent. Calibrated models for each facility, as described in Chapter 4.0, were used to run simulations for each facility using a matrix of PSD values (represented by various removal rates) and initial chlorine concentrations. The intra-particle chlorine concentrations generated by the model in simulations were used to estimate the concentration of target microorganisms in plant effluent particles under a variety of PSD and chlorine concentration conditions. The estimated effluent concentrations of occluded target microorganisms were compared with the treatment standards that formed the bases for the particle guidelines. The end result was a range of PSD and chlorine concentration combinations that resulted in acceptable treatment.

## 5.2 Model Simulations

Following calibration of the model for each facility, model simulations were run using different scenarios of initial chlorine concentration and PSD to determine the resulting intra-particle chlorine concentrations, which were used to estimate the concentration of occluded total coliform bacteria. The calibrated model parameters were used for all simulations (Table 4-3). As stated in Chapter 4.0, model calibration was performed using initial chlorine concentrations that were lower than the initial chlorine concentrations in the disinfection experiments. The differences accounted for the instantaneous chlorine demand that occurs prior to the first measurement of chlorine residual. The instantaneous chlorine demand was assumed constant for the conditions of each simulation. Model simulations were run using the initial chlorine concentrations shown in Table 5-1.

**Table 5-1. Initial Chlorine Concentrations Used in Experiments and in Model Simulations.**

| Parameter | Facility A | Facility B | Facility C | Facility D | Facility E | Facility F | Facility G |
|---|---|---|---|---|---|---|---|
| Initial chlorine concentration added in the disinfection experiment (mg/L) | 8 | 6 | 6 | 4 | 20 | 12 | 1.5 |
| Instantaneous chlorine demand (mg/L) | 3.1 | 3.1 | 2.0 | 0.6 | 0.0 | 0.8 | 0.5 |
| Initial chlorine concentrations for simulations (mg/L) | 0.9<br>1.9<br>2.9<br>4.9<br>6.9<br>8.9<br>10.9<br>12.9<br>14.9<br>16.9<br>26.9<br>36.9<br>46.9<br>56.9<br>66.9<br>76.9<br>86.9<br>96.9 | 0.9<br>2.9<br>4.9<br>6.9<br>8.9<br>10.9<br>12.9<br>14.9<br>24.9<br>34.9<br>44.9<br>54.9<br>64.9<br>74.9<br>84.9<br>94.9 | 2<br>4<br>6<br>8<br>10<br>12<br>14<br>16<br>26<br>36<br>46<br>56 | 1.4<br>3.4<br>5.4<br>7.4<br>9.4<br>11.4<br>13.4<br>15.4<br>25.4<br>35.4<br>45.4<br>55.4<br>65.4<br>75.4<br>85.4<br>95.4 | 1.6<br>3.6<br>5.6<br>7.6<br>9.6<br>11.6<br>13.6<br>15.6<br>17.6<br>19.6<br>21.6<br>23.6<br>33.6<br>43.6<br>53.6<br>63.6 | 1.2<br>3.2<br>5.2<br>7.2<br>9.2<br>11.2<br>13.2<br>15.2<br>17.2<br>19.2<br>21.2<br>23.2<br>33.2<br>43.2<br>53.2<br>63.2 | 1<br>2<br>3<br>4<br>5<br>7<br>9<br>11<br>21<br>31<br>41<br>51 |

The following filtration scenarios were considered for each facility:

- No filtration,
- 30% removal of particles in all particle size classes,
- 60% removal of particles in all particle size classes, and
- 90% removal of particles in all particle size classes.

The above-listed particle conditions for simulations were based on the assumption of consistent particle removal for all particle sizes, such that large particles are removed at the same rate as small particles. At least fifteen model simulations were run (in addition to the initial model run completed for calibration) for each facility, each representing a unique combination of initial chlorine concentration and PSD. Additional simulations were run to determine conditions that resulted in final total coliform concentrations that were below the treatment objective.

## 5.3 Optimization Protocols Based on Indicator Organisms

### 5.3.1 Particle Guidelines Based on Indicator Organisms

The particle guidelines shown in Table 2-3 were developed based on assumptions that the frequency of association and the microbial density are both one (F=1 and P=1). The guidelines were updated to reflect the frequency of association that was measured for total coliform bacteria at each facility (Table 3-2) as shown in Table 5-2. The low frequency of association values observed in this study result in particle guidelines that are high, much higher than the regulatory limits for indicator organisms shown in Table 2-2 (e.g. total coliform limits of 2.2-200 MPN/100 mL). A conservative approach would be to assume that the frequency of association is one, and to use the most stringent effluent requirement for indicator organisms, which is 2 *E. coli* or coliform bacteria (total or fecal) per 100 mL. The value of P was not updated and was kept at one.

**Table 5-2. Particle Guidelines (# / 100 mL) Based on Indicator Organisms, with Measured Particle Concentrations in Secondary Effluent Samples Shown for Comparison.[1]**

| Microbe | Exposure route | Facility A | Facility B | Facility C | Facility D | Facility E | Facility F | Facility G |
|---|---|---|---|---|---|---|---|---|
| *E. coli* and coliform bacteria | Drinking water and reclaimed water | 490 | 650 | 14,000 | 230 | 960 | 36,000 | 150 |
| Enterococci | Recreational fresh water | 8100 | 11,000 | 230,000 | 3800 | 16,000 | 580,000 | 2400 |
| Enterococci | Recreational marine water | 8600 | 12,000 | 250,000 | 4100 | 17,000 | 620,000 | 2500 |
| Measured particle (>11 μm) concentrations (#/100 mL) | | **4900** | **3000** | **160,000** | **81,000** | **82,000** | **63,000** | **150,000** |

[1]Particle guidelines were calculated using Equation 2-1 with measured values of F and P=1. Particle guidelines represent effluent particles limits that ensure regulatory requirements are met for the indicated microorganisms, assuming that the dispersed target microorganisms are completely inactivated and that the occluded target microorganisms are not inactivated in the disinfection system.

The particle concentrations observed for particles larger than 11 μm are shown along with the particle guidelines in Table 5-2. As indicated, the particle guidelines were met for recreational exposure scenarios for Facilities A, B, C, and F. The particle guidelines were not met for any of the facilities when drinking water or reclaimed water was the exposure route. None of the particle guidelines were met for Facilities D, E and G.

### 5.3.2 Estimated Concentration of Occluded Total Coliform Following Disinfection

Equation 1-2, in combination with the model output and the measured inactivation rate coefficients, was used to predict the concentration of occluded total coliform bacteria that would survive a 90-minute disinfection period (Table 5-3). The total coliform estimates shown in Table 5-3 were calculated assuming that the total coliform bacteria were occluded in the largest particles. An example of this preferential association with large particles is that if there were only 20 particles with occluded total coliform bacteria in a disinfection reactor, these 20 particles will be the largest particles. In the preliminary modeling efforts additional calculations were made assuming that the occlusion of total coliform bacteria in particles was evenly distributed throughout all of the particle size classes. For example, if 1% of particles had occluded total coliform bacteria, then the particles in each particle size class will have a frequency of association of 1%. However, the assumption that occluded total coliform bacteria were evenly distributed throughout the different particles size classes yielded results that were less conservative than those obtained assuming that occluded total coliform bacteria inhabited the largest particles. The assumption regarding total coliform occlusion in particles affects the model calibration and separate calibration procedures must be performed when different assumptions are made. For this study, a single assumption regarding total coliform occlusion was made and the model was only calibrated once for each facility.

In Table 5-3 the shaded numbers represent simulations where the calculated total coliform concentrations were lower than the regulatory standard of 2 MPN/100 mL. In the table only the treatment conditions (chlorine concentration and particle removal rates) that result in final total coliform concentrations near the treatment objective of 2 MPN/100 mL are shown. In most cases, removing most of the particles (90% removal) was highly effective for meeting the treatment objective. Where lower particle removal is applied (30-60% removal), less chlorine is required than if no particle removal is practiced.

The calculated estimates for total coliform bacteria (Table 5-3) were plotted as contour plots (Figures 5-1). The information used to develop the contour plots was based on the preferential association of total coliform bacteria with larger particles. One way to interpret the information shown on Figures 5-1 is to evaluate the effects of constraints at the treatment facilities. For example, if the manager for Facility A (Figure 5-1) determines that the filtration system is only reliably reducing particle concentrations by 50%, the manager knows that approximately 10 mg/L of chlorine must be added in the disinfection system to sufficiently reduce the concentration of total coliform bacteria to concentrations below 2 MPN/100 mL. Similarly, managers of treatment systems may want to constrain the quantity of chlorine added for economic reasons. Particle removal has a strong impact on the amount of chlorine required for Facility E. If 90% removal of particles is possible, the facility managers can reduce the amount of chlorine required from 19.6 mg/L to 3.6 mg/L, according to the results of the modeling efforts.

Table 5-3. Occluded Total Coliform Bacteria (MPN/100 mL) Surviving a 90-Minute Contact Time, as Calculated Using Model Output and Assuming the Frequency of Association Is Higher for Larger Particles.[1,2]

| | Initial chlorine conc. (mg/L)[3] | Alterations to the PSD | | | |
|---|---|---|---|---|---|
| | | No change | 30% removal | 60% removal | 90% removal |
| **Facility A** | 4 | 14.3 | 10.0 | 5.7 | 1.4 |
| | 5 | 9.9 | 6.9 | 3.9 | 1.0 |
| | 6 | 6.8 | 4.8 | 2.7 | 0.7 |
| | 8 | 3.2 | 2.3 | 1.3 | 0.3 |
| | 10 | 1.5 | 1.1 | 0.6 | 0.2 |
| **Facility B** | 4 | 5.1 | 3.7 | 2.2 | 0.4 |
| | 6 | 0.5 | 0.4 | 0.2 | 0.0 |
| **Facility C** | 4 | 5.7 | 5.7 | 4.0 | 1.0 |
| | 6 | 1.9 | 1.9 | 1.3 | 0.3 |
| **Facility D** | 2 | 18.8 | 13.3 | 7.7 | 2.2 |
| | 4 | 4.6 | 3.2 | 1.8 | 0.4 |
| | 6 | 1.6 | 1.1 | 0.6 | 0.1 |
| **Facility E** | 1.6 | 41.5 | 14.3 | 7.8 | 2.6 |
| | 3.6 | 24.2 | 6.5 | 3.5 | 1.3 |
| | 5.6 | 15.8 | 3.3 | 1.7 | 0.7 |
| | 7.6 | 11.0 | 1.9 | 1.0 | 0.4 |
| | 9.6 | 8.0 | 1.3 | 0.6 | 0.3 |
| | 11.6 | 5.9 | 0.9 | 0.4 | 0.2 |
| | 13.6 | 4.5 | 0.6 | 0.3 | 0.2 |
| | 15.6 | 3.4 | 0.5 | 0.2 | 0.1 |
| | 17.6 | 2.5 | 0.4 | 0.2 | 0.1 |
| | 19.6 | 1.9 | 0.3 | 0.1 | 0.1 |
| **Facility F** | 2 | 13.0 | 12.4 | 7.1 | 1.8 |
| | 4 | 5.4 | 5.2 | 2.9 | 0.7 |
| | 6 | 2.3 | 2.1 | 1.2 | 0.3 |
| | 8 | 0.9 | 0.9 | 0.5 | 0.1 |
| **Facility G** | 4.5 | 34.1 | 23.8 | 13.5 | 3.4 |
| | 5.5 | 15.6 | 10.8 | 6.2 | 1.5 |
| | 7.5 | 4.4 | 3.1 | 1.7 | 0.4 |
| | 9.5 | 1.4 | 1.0 | 0.6 | 0.1 |

[1]Shaded numbers represent concentrations where the particle guideline of 2/100 mL was met.
[2]Estimates were calculated assuming total coliform occluded in larger particles prior to smaller particles.
[3]Initial chlorine concentration is the initial model concentration plus the instantaneous chlorine demand.

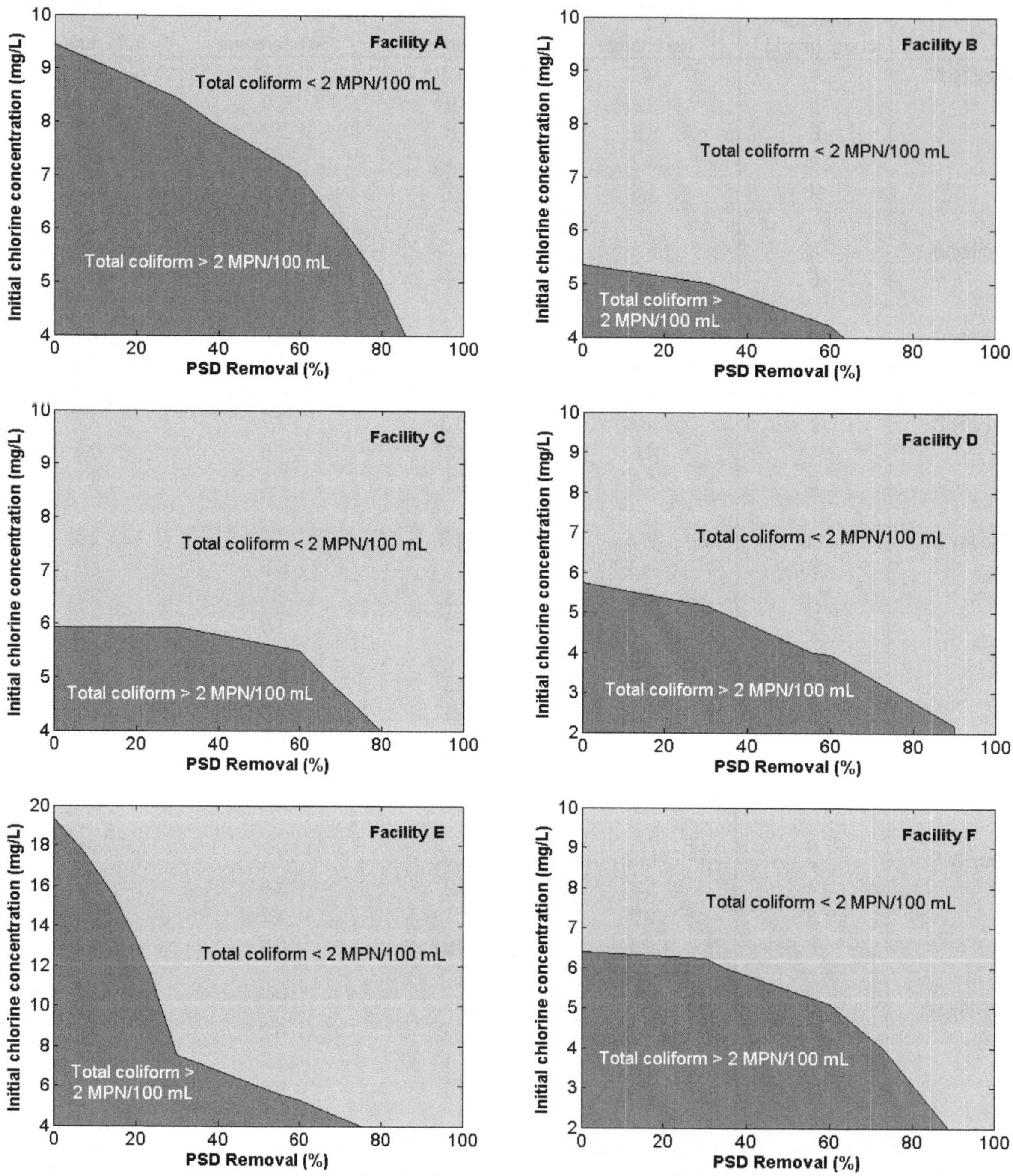


**Figure 5-1. Combinations of Particle Removal and Initial Total Chlorine Concentration That Result in Occluded Total Coliform Concentrations Greater Than and Less Than 2 MPN/100 mL, as Predicted Using the Model Output.**

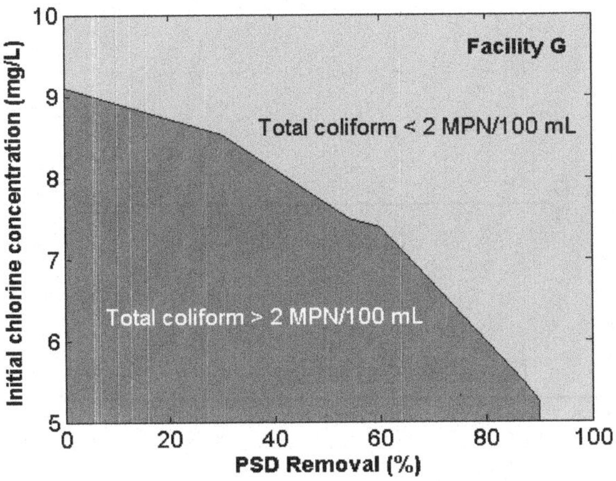


**Figure 5-1 continued. Combinations of Particle Removal and Initial Total Chlorine Concentration That Result in Occluded Total Coliform Concentrations Greater Than and Less Than 2 MPN/100 mL, as Predicted Using the Model Output.**

The chlorine concentrations shown on the contour plots in Figure 5-1 refer to the initial chlorine that would be required, which includes the instantaneous chlorine demand. One of the assumptions in this study is that the instantaneous chlorine demand does not change when different amount of chlorine are added initially. Further study could be done to verify this assumption or to quantify the relationship between initial chlorine added and the instantaneous chlorine demand. The instantaneous chlorine demand is subtracted when conducting the model simulations because it is assumed that this portion of the chlorine added is reacting with soluble organic matter and is not transferring into particles.

By interpolating between the values shown in Table 5-3, combinations of initial chlorine concentration and particle removal were identified that resulted in total coliform concentrations of 2 MPN/100 mL. The results of the interpolation exercise are shown in Figure 5-2 for facilities with data that spanned values less than and greater than 2 MPN/100 mL. Again, the total coliform concentrations used to generate Figure 5-2 were calculated using Equation 1-2 with 90-minutes of contact time and intra-particle chlorine concentrations from model simulations (Table 5-3). The lines in Figure 5-2 represent a best fit for the data points. One difference between the information shown in Table 5-3 and in Figure 5-2 is that the particle removal rates were converted to particle counts.

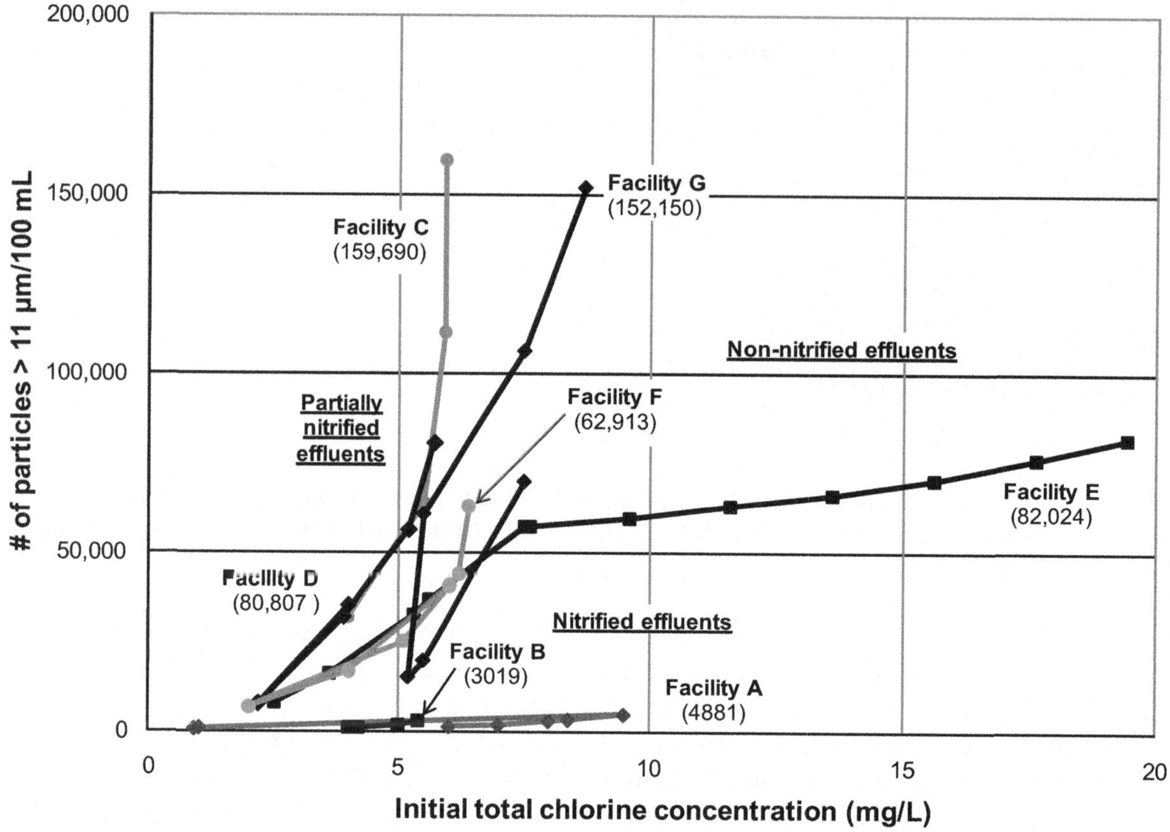


**Figure 5-2. Particle and Initial Total Chlorine Concentrations That Result in Occluded Total Coliform Concentrations of 2 MPN/100 mL (Original Particle Counts Are Listed in Parentheses).**

As shown in Figure 5-2, the nitrified facilities, Facility A and Facility B, have low particle counts and the slopes of the lines are nearly flat. For these facilities, small decreases in the particle concentrations would result in large decreases in the amount of chlorine required to achieve the treatment objective of 2 MPN/100 mL. The partially-nitrified effluents, Facility C and Facility D, have lines with much higher slopes, indicating that there is a narrow range of chlorine concentrations required regardless of particle removal that can be provided. The non-nitrified effluents, Facilities E and Facility G, have steep slopes that are similar to those for the partially-nitrified effluents, but the slope of the line for Facility E is much lower and more similar to the slopes of the lines observed for Facilities A and B. For Facility E decreases in particle counts will have a strong influence on the amount of chlorine required to meet the treatment objective of 2 MPN/100 mL.

## 5.4 Discussion and Conclusions

Following model calibration, model simulations were completed for a variety of particle removal rates and initial chlorine concentrations. Initial chlorine concentrations used in the simulations were assumed not to include the instantaneous chlorine demand, which was assumed constant and not affected by the initial chlorine concentration added to the reactors. Particle concentrations that were used were a fraction of the initial particle concentrations: 0% removal, 30% removal, 60% removal, and 90% removal. Consistent particle removal was assumed for each particle size class. Use of removal rates for particle concentrations was intended to simulate removal of particles in filtration systems.

The particle guidelines were updated from the generic format shown in Chapter 2.0, where F and P were both assumed as one, to particle guidelines that were based on values of F observed in the laboratory studies. The value of P was maintained at one. Updating the particle guidelines based on measured values of F may cause the guidelines to be more reasonable. However, it would be more conservative to use particle guidelines where F is assumed to be one. Using the measured values of F ignores the stochastic nature of the frequency of microorganisms in particles; F is a variable with a probabilistic range of expected values. The frequency of association was only measured once for each of the facilities during this study. If, on a given day, the frequency of association were much higher than what was measured in this study, the concentration of occluded microorganisms could be much higher than expected. For example if 1% of particles contain target microorganisms and the allowable quantity of the target microorganisms in the effluent is one in 100 mL, then the particle guideline would be 100/100 mL. However, for any given random sample of 100 particles there is a probability that the number of target microorganisms will be higher than one. Assuming F=1 eliminates the uncertainty surrounding the probability that a given sample will have a higher frequency of occluded microorganisms than the average frequency.

Assuming a value of one for P may not be the most conservative approach. However, it is consistent with the current analytical techniques and with the existing regulatory framework. In culture-based microbiological tests, a microorganism occluded in a particle is counted as a single colony (bacterial plate count) or plaque (viral plate count) or positive tube (multiple-tube fermentation test). If a particle contains more than one microorganism, it is still counted as a single microorganism in the culture-based tests and there is no established method to reliably extract all microorganisms prior to enumeration. Established regulations that are based on culture-based microbiological methods do not distinguish between a single occluded target microorganisms and a particle harboring multiple target microorganisms. The health risks of particles containing multiple microorganisms are also not documented. Given the more conservative nature of one for F and the uncertainty of numbers larger than one for P, development of particle guidelines using F=1 and P=1 is recommended at this time. Basically, by assuming that F=1 the approach is sufficiently conservative because the system is designed to inactivate microorganisms from all particles of all sizes so the density of microorganisms within particles is immaterial.

The model simulations yielded intra-particle chlorine concentrations for various radial distances within the particles as well as the concentration of occluded total coliform bacteria over time. Total coliform bacteria were assumed to inhabit the center of particles. The justification of this approach is that if the center of the particles were adequately saturated with the chlorine, all total coliform bacteria potentially present in the particle will be inactivated. The inactivation coefficient used was determined experimentally for filtered (2 μm) samples, which is based on the assumption that occluded target microorganisms have the same inactivation characteristics as dispersed target microorganisms and that only diffusion of the chlorine through the particles is limiting inactivation.

The calculated particle guidelines were compared with particle data collected in the laboratory studies. Although the particle guidelines were met in some cases, the observed particle counts were typically higher than the particle guidelines. Where the particle guidelines are not met, the disinfection system must be sufficiently designed and operated to inactivate occluded target microorganisms. The particle guidelines shown were based on meeting the regulatory limits for total coliform bacteria occluded in particles and do not address inactivation of dispersed microorganisms. Appropriate chlorine doses must be selected and applied to inactivate dispersed microorganisms as well as the occluded microorganisms.

The final occluded total coliform concentrations obtained from the model simulations is contained in tabular form in Table 5-3 and in the contour plots in Figures 5-1. For each facility, combinations of initial chlorine concentration and particle removal are identified that result in total coliform concentrations relative to the selected standard of 2 MPN/100 mL. For a particular facility, managers can identify appropriate chlorine concentrations for a given particle removal rate. Conversely, target particle removal rates can be identified if the chlorine concentration is constrained by facility conditions or chemical costs. The values presented in Table 5-3 as well as in Figures 5-1 are based on a contact time of 90-minutes. It is possible to generate similar tables and figures using a different contact time, which would require subsequent calculations using model output for a different time step. Similarly, alternative treatment objectives could be used and would result in different contour lines for Figures 5-1; the shaded numbers in Table 5-3 would also change.

The results for Facility G are based on preliminary estimates, as insufficient data are available from the disinfection tests to adequately calibrate the model for this facility. In the disinfection test the tailing portion of the inactivation curve is higher than the concentration of particles in the sample. The remainder of the data set is sufficient for model calibration and simulation exercises. As a result, an assumption regarding one of the model parameters was made (Chapter 4.0) and the calibration and simulation exercises were completed as an estimate of the results that would be expected from this facility. Additional experimental work would be required to identify data to complete the model calibration.

As shown in Figure 5-2, particle counts for the facilities vary widely. In particular, the facilities with non-nitrified effluents have high particle counts, although Facility D has a partially nitrified effluent (ammonia of approximately 1 mg/L) and it also has a high particle count. The MCRT information provided by the facilities is useful for interpreting the results. Facility A has a very high MCRT of 40 days and a low total particle count. While the MCRT of Facility B is half that of Facility A (20 days), the particle count of this facility is also low. The MCRT of Facility D is much lower (5 days), which explains the presence of ammonia in the effluent as well as the higher particle count. Facilities F and G produce non-nitrified effluents (MCRT of less than 5 days at each facility) and have high particle counts.

A comparison of the slopes of the best-fit line through the data points indicates that the operating conditions at the facilities impact the disinfection characteristics. For the facilities with a flatter slope, such as Facilities A, B, and E, a small change in the particle concentration results in a substantial reduction of chlorine required. For the other facilities, the slopes are much higher and a larger reduction in the particle concentration is required to achieve a similar reduction in chlorine required to meet the treatment objective of 2 MPN/100 mL. The trends apparent in Figure 5-2 also demonstrate the site-specific nature of chlorine demand and inactivation characteristics, requiring investigation of particle and disinfection characteristics at individual treatment facilities.

The information shown in Figure 5-2 is useful for managers of treatment facilities because the effects of operational changes (particle reduction through filtration or changes in chlorine applied) can be evaluated. For example, if a proposed filtration system is capable of reducing the total particle concentration by 50%, the manager can estimate the resulting decrease in total chlorine needed to inactivate the occluded total coliform bacteria. Of course, disinfection system operating conditions must also be selected such that the dispersed target microorganisms are inactivated as well as the occluded ones. All of the information in Figure 5-2 was obtained assuming that the contact time in the chlorine reactor was 90-minutes. Similar plots could be developed with a different contact time.

CHAPTER 6.0

# FUTURE WORK TO EXTEND APPLICATION OF OPTIMIZATION PROTOCOLS

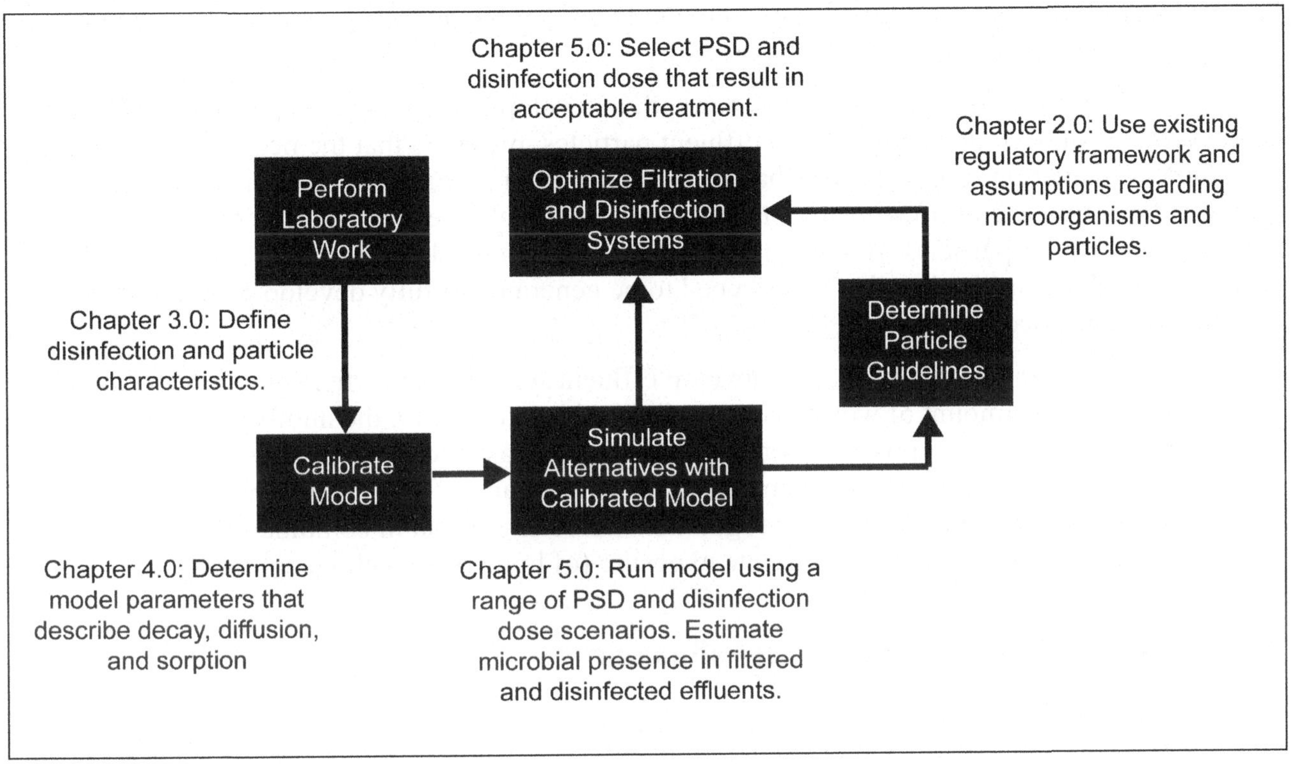


## 6.1 Introduction

Ideally, epidemiological studies that characterize health risk associated with varying levels of exposure to selected constituents would be used to develop water quality standards. However, epidemiological studies are expensive, time-consuming, difficult to conduct, and often not practical. In epidemiological studies, it is difficult to isolate health effects and determine the cause and effect of different factors. For example, it may be difficult to determine the source of illness among swimmers using a water body that contains reclaimed water. In the absence of epidemiological studies for particles in secondary and tertiary effluents, a quantitative microbial risk assessment approach was used to determine acceptable concentration of effluent particles. In this health-based approach, the probabilistic incidence of illness was correlated with possible exposure of humans to particle-associated pathogens.

Risk assessments are often used in lieu of epidemiological studies. Health data are collected, generalized and used in calculations of the estimated risk. One of the basic concepts in risk assessment is the dose-response model, which defines the range of expected responses (for example, illness or death) to a stressor such as a toxic chemical or in this case, a pathogenic microorganism. Often, response data are collected for high doses (such as in an outbreak) and extrapolated to responses expected for low doses that are much more difficult to observe. To perform a quantitative microbial risk assessment, data are collected to estimate the exposure that individuals may have to a pathogen through specific exposure routes. Risk assessment is an empirical approach, resulting in information based on experience; published health information is used to determine probable risk. The information gleaned from risk assessments contains a large degree of uncertainty as it is based on several factors, each of which contains uncertainty. Despite the uncertainty involving risk assessments, they do have the advantage of incorporating health information into standards without the necessity of conducting epidemiological studies (which also contain a large degree of uncertainty).

The optimization approach could be used to develop pathogen-based particle guidelines and estimates of occluded pathogens in effluent particles, provided that the necessary data were available. Calculations are presented that are based on assumptions for pathogens, and these calculations provide a best-estimate of the risks posed by pathogens potentially present in effluent wastewater particles. A summary of future research needs is presented at the end of the chapter, which describes data that would need to be generated to fully develop pathogen-based optimization protocols.

Four constituents are needed to develop effluent wastewater particle guidelines with a pathogen basis: the amount of water that an individual is exposed to, the number of pathogens that an individual can be exposed to without incurring unreasonable health risks, and the frequency and density of microorganisms in particles. The acceptable level of exposure to pathogens is determined using existing regulations and risk criteria in combination with published dose-response models that relate exposure (of humans to pathogens) to an end response such as illness or infection.

## 6.2 Particle Guidelines Based on Pathogens

### 6.2.1 Exposure Scenarios

The exposure scenarios and assumptions used to develop particle guidelines based on pathogens are the same as those used to develop particle guidelines based on indicator organisms (Section 2.2).

### 6.2.2 Risk Criteria, Microbial Standards, and Treatment Standards

Regulatory guidelines that ensure the microbiological safety of environmental waters are based on limits for specific microorganisms (often corresponding with detection limits), the capability of treatment technologies and trains to provide a target log removal of microorganisms, and criteria that are used in risk assessments. The following is a summary of existing regulatory guidelines that were the bases of the particle guidelines, representing d or $d/V_c$ in Equation 2-1.

The U.S. EPA has used the following risk criteria for swimming-related gastroenteritis resulting from recreational exposure to wastewater effluents discharged into surface waters: eight cases of illness occurring for every 1,000 swimmers at fresh water beaches, and 19 cases of illness for every 1,000 swimmers at marine beaches (U.S. EPA, 1986). The U.S. EPA criteria were developed as the result of epidemiological studies. Concentrations of indicator organisms corresponding to this risk criterion are indicated as Guideline 1 in Table 2-2.

In addition to the total coliform limits in the California Recycled Water Criteria, production of disinfected tertiary recycled water requires a total of 5-log removal of viruses in the filtration and disinfection systems (CA DPH, 2001). The 5-log removal standard was developed based on results of a pilot study performed by the County Sanitation Districts of L.A. County where poliovirus was spiked into the influent and its removal was observed in parallel tertiary treatment trains (CSDLAC, 1977). Demonstration of 5-log removal of viruses must be performed for disinfectants other than chlorine and for filtration systems, and can be performed using F-specific bacteriophage MS2, poliovirus, or other equivalently resistant viruses (CA DPH, 2008).

To apply the treatment-based standard of 5-log removal of viruses in filtration and disinfection systems for reclaimed water, information must be obtained regarding the typical presence of viruses in secondary effluent. The occurrence of pathogens in wastewater is variable depending on the geographical location, climate, land use, and demographics. In one study where pathogens were monitored in various points within the treatment trains at six wastewater treatment facilities, 22 of 32 samples had enterovirus concentrations that were above the detection limits and the range of data for the positive samples was 0.0005 to 3.3 MPN/100 mL (Rose et al., 2004). Applying the standard of 5-log removal, the effluents in the study by Rose et al. (2004) would be expected to have concentrations of $5x10^{-9}$ to $3.3x10^{-5}$ MPN/100 mL enteroviruses in the filtered and disinfected effluent.

It is not clear how typical virus concentrations would vary depending on the status of public health and rates of illness within a community. Virus concentrations have been shown to vary seasonally (Hejkal et al., 1984; Irving and Smith, 1981). Hejkal et al. (1984) found that rotavirus concentrations were higher during the winter and spring, but that there was no seasonal variation of enterovirus concentrations. Irving and Smith (1981) found that total virus concentrations were highest during the winter. There are insufficient data to suggest that the virus concentrations reported by Rose et al. (2004) should be increased to account for higher numbers that would have been observed had the study been conducted in a different region, at a different time of year, or in a different community.

The National Primary Drinking Water Regulations include regulatory limits of zero for the following indicator organisms and pathogens: total coliforms, fecal coliforms, *Escherichia coli*, *Cryptosporidium*, *Giardia lamblia*, *Legionella*, and viruses (U.S. EPA, 2002). The target of zero refers to a value of zero in any volume of water tested; therefore, the detection limit of each test is a function of the volume tested. Although specific to drinking water, microbial drinking water standards may be appropriate for reclaimed water projects.

For utilities providing potable water by treating surface water or ground water under the direct influence of surface water must meet the following treatment standards: 2-log removal of *Cryptosporidium* (U.S. EPA, 2002), 3-log removal of *Giardia lamblia* (U.S. EPA, 1989), and 4-log removal of viruses (U.S. EPA, 1989). In addition, public water systems using filtration that have high concentrations of *Cryptosporidium* in their source water are required to provide an additional 1.0 to 2.5-log removal of *Cryptosporidium* (U.S. EPA, 2006). To understand the acceptable levels of pathogens in drinking water, the treatment standards must be reviewed in combination with typical levels of these pathogens in source waters. Values as high as 484 oocysts/L for *Cryptosporidium* and 66 cysts/L for *Giardia* were located for source waters used to produce potable water supplies (Le Chevallier et al., 1991). The maximum value for enteroviruses reported by Payment et al. (2000) was 62.3 MPNIU/L. The values used here represent the maximum values reported in studies; typical values of pathogens are lower and often below detection limits. Applying the above treatment standards to the values reported by LeChevallier et al. (1991) and Payment et al. (2000), the concentrations in the finished water would be 0.153 oocysts/100 mL, 0.66 cysts/100 mL, and 0.0623 MPNIU/100 mL.

As part of the initial Surface Water Treatment Rule, the U.S. EPA provided guidance for acceptable risk of exposure to microorganisms in potable water supplies: "EPA believes that providing treatment to ensure less than one case of microbiologically caused illness per year per 10,000 people is a reasonable goal" (U.S. EPA, 1989, p. 27490). Although the rule is specific to exposure to *Giardia lamblia* in treated surface water that is used as a potable water source, the statement made by the U.S. EPA is general and has been used for exposure to other pathogens (e.g. viruses) and other types of exposure scenarios (e.g. groundwater). The U.S. EPA drinking water criterion has also been applied to water reuse projects where a volume of water is assumed to be accidentally ingested (e.g., Tanaka et al., 1998; York and Walker-Coleman, 2000).

The State of Florida Department of Environmental Protection has developed risk-based guidelines for *Giardia* and *Cryptosporidium*, although these guidelines are not regulatory and are not enforceable (York and Walker-Coleman, 2000). The State has instead opted to collect data from utilities with water reuse programs to have a better understanding of the presence of *Giardia* and *Cryptosporidium* in reclaimed water. A summary of the Florida guidelines is shown in Table 6-1. The average limits are based on a 1:10,000 risk of infection for ingestion of 1 mL of reclaimed water daily for an entire year. The maximum limits are based on a 1:10,000 risk of infection for a single ingestion of 100 mL of reclaimed water.

**Table 6-1. Pathogen Guidelines for Reclaimed Water in Florida (York and Walker-Coleman, 2000).**

| Organism | Units | Average limit | Maximum limit |
|---|---|---|---|
| Enterovirus | PFU/100 L | 0.044 | 0.165 |
| *Cryptosporidium* | Viable cysts/100 L | 5.8 | 22 |
| *Giardia* | Viable cysts/100 L | 1.4 | 5.0 |

In summary, there are numerous methods for determining acceptable particle concentrations with a pathogen basis as is shown in Table 6-2. Guideline 1 was based on risk assessment calculations. Guideline 2 was based on the capabilities of available analytical methods. Guidelines 3 and 4 were based on the capabilities of available treatment technologies. Guidelines 5 and 6 are risk criteria that are used to perform risk assessments to develop appropriate effluent limits, as is done in the following section. Guidelines 5 and 6 were developed based on epidemiological studies.

Table 6-2. Regulatory Pathogen Guidelines That Were the Bases for Particle Guidelines.

| | Origin of recommendation | Recommendation | Other information needed |
|---|---|---|---|
| 1 | FL DEP | Reclaimed water: enterovirus=0.044 PFU/100 L, *Cryptosporidium* =5.8 viable oocysts /100 L, and *Giardia*=1.4 viable cysts/100 L | No |
| 2 | EPA | Potable water: detection limits for *Cryptosporidium, Giardia lamblia, Legionella*, and viruses | No |
| 3 | EPA | Treatment of potable water: 2-log removal of *Cryptosporidium*, 3-log removal of *Giardia lamblia*, and 4-log removal of viruses plus an additional 1.0-2.5-log removal of *Cryptosporidium* for source waters with elevated concentrations. | data for surface water sources |
| 4 | CA DPH | Reclaimed water: 5-log removal of viruses using a combination of filtration and disinfection for wastewater that has been oxidized | data for viral concentrations in oxidized wastewater |
| 5 | EPA | Potable water: 1 infection annually in a population of 10,000 | dose-response data[1] |
| 6 | EPA | Recreational water: 8 cases of illness occurring for every 1,000 swimmers at fresh water beaches, and 19 cases of illness for every 1,000 swimmers at marine beaches | dose-response data |

[1]In Table 6-2 dose-response refers to information that relates an ingested quantity of a pathogen (dose) to a health endpoint such as illness or infection (response).

### 6.2.3 Dose-Response Models for Exposure to Pathogens

To develop particle guidelines based on pathogens using Recommendations 5 and 6 from Table 6-2, dose-response data from the field of risk assessment are needed. Where the acceptable incidence of infection or illness is defined, the corresponding acceptable exposure (with units of pathogen number (e.g., CFU)) to a particular pathogen can be calculated using published dose-response models (Haas et al., 1999; Loge et al., 2002). The following pathogens were used for the risk assessment based on their relevance as waterborne pathogens and the availability of published dose-response data: *Escherichia coli* ETEC, *Escherichia coli* O157:H7, *Shigella* spp., rotavirus, *Salmonella* spp. *Cryptosporidium parvum*, and *Giardia lamblia*. The beta-Poisson dose-response model, as described by Equation 6-1, was used for bacteria and viruses, and the exponential dose-response model, as described by Equation 6-2, was used for protozoan pathogens (Haas et al., 1999).

$$P_{bp}(d) = 1 - \left(1 + \frac{d}{N_{50}}\left(2^{1/\alpha} - 1\right)\right)^{-\alpha} \qquad (6\text{-}1)$$

$$P_e(d) = 1 - e^{-r_e d} \qquad (6\text{-}2)$$

The probability of infection or illness ($P_{bp}$ and $P_e$) for various exposures corresponds with the appropriate risk criterion; it is dependent on the dose (d) of the pathogen ingested, which in this case is the total number of organisms ingested in a single event. For exposure to drinking water the probability of infection is converted to a daily risk. In the case of reclaimed water ingested during swimming, one day of swimming is considered one event. The value of $N_{50}$ is the median infectious dose, which is the number of a specific type of organism that must be ingested to cause infection in half of the exposed population. The variables $\alpha$ and $r_e$ are model parameters that are specific to individual pathogens; published values are available for these parameters. Table 6-3 contains a summary of dose-response data that were used in calculations. To convert the probability of infection to a probability of illness, the probability of infection was multiplied by a morbidity ratio. Data regarding the incidence of illness relative to the incidence of infection were not located for all pathogens. However, risk assessments can be performed using various health endpoints such as infection or illness. Where data regarding the morbidity ratio were available, illness was the selected health endpoint. For pathogens where the morbidity ratio was unknown, infection was selected as the health endpoint.

**Table 6-3. Dose-Response Data for Human Exposure to Pathogens.**

| Organism | Dose/response parameters (ref.) | Morbidity ratio (ref.) |
|---|---|---|
| *E. coli* O157:H7 | $\alpha$ = 0.21; $N_{50}$ = 1120 (Haas et al., 1999) | |
| *E. coli* ETEC | $\alpha$ = 0.18; $N_{50}$ = $8.6*10^7$ (Haas et al., 1999) | |
| *Shigella* | $\alpha$ = 0.21; $N_{50}$ = 1120 (Haas et al., 1999) | |
| *Salmonella* | $\alpha$ = 0.31; $N_{50}$ = $2.36*10^4$ (Haas et al., 1999) | 0.22 (McCullough and Eisele, 1951) |
| Rotavirus | $\alpha$ = 0.265; $N_{50}$ = 5.6 (Haas et al., 1993) | 0.57 (Ward et al., 1986) |
| *Giardia lamblia* | $r_e$ = 0.02 (Rose et al., 1991) | 0.5 (Rose et al., 1991) |
| *Cryptosporidium parvum* | $r_e$ = 0.0042 (Haas et al., 1996) | 0.39 (Dupont et al., 1995) |

### 6.2.4 Acceptable Particle Concentrations Based on Pathogens

The dose-response data in Table 6-3 was used in combination with Equations 6-1 and 6-2 to estimate a microbial dose corresponding to the probability of infection or illness. The estimated dose for each pathogen was then used to calculate the allowable concentration of particles using Equation 2-1, which would not exceed the risk-based criterion. The calculated particle concentrations based on pathogens are shown in Appendix A. Similar to particle guidelines that were developed with an indicator organism basis (Table 2-3), particle concentrations developed using a pathogen basis are relevant to particles that have a diameter larger than 11 μm, the lower limit of particle size that can be quantified using the method employed in this study. Three values of F were considered (0.0044, 0.24, and 1) in calculations using Equation 2-1 and P was always assumed to be 1. The acceptable level of risk for drinking water defined by the U.S. EPA (Guideline 5 in Table 6-2) is an annual risk, so it was converted to a daily risk prior to use in the dose-response model ($P_a = 1-(1-P_s)^N$ where $P_s$ is the probability of infection or illness based on a single exposure (e.g. one day) and N is the number of exposures per year (365) and $P_a$ is the annual probability). Values of $P_s$ were substituted for $P_{bp}$ and $P_e$ in Equations 6-1 and 6-2.

As shown Table A-1 in Appendix A, there are multiple ways to calculate particle guidelines using a risk assessment framework, and the resulting particle guidelines are a function of the exposure route, microorganism, and the microbial frequency and density in particles. The particle guidelines based on indicator organisms (Chapter 2.0) and pathogens are shown in Table A-2 in Appendix A. The lowest particle guideline is $5.0 \times 10^{-9}$ particles/100 mL, which is equivalent to one particle in 20 million liters. Calculation of the lowest particle guideline was based on Guideline 4 in Table 6-2, which is the guideline for 5-log removal of viruses in reclaimed water. The second lowest particle guideline ($2.6 \times 10^{-8}$ particles larger than 11 μm/100 mL) was calculated for exposure to Rotavirus in drinking water, which was calculated using the risk assessment framework. The highest particle guideline for pathogens ($1.6 \times 10^{-5}$ particles larger than 11 μm/100 mL) was calculated based on exposure to pathogenic *E. coli* ETEC in recreational marine waters. Particle guidelines were not calculated for pathogens using Guideline 2 in Table 6-2 because detection limits for the various pathogens is dependent on sample volume and assumptions would be necessary to perform further calculations.

The particle guidelines shown in Tables A-1 and A-2 of the Appendix A were calculated based on assumptions for the frequency of occurrence of microorganisms in particles and of the microbial density. Site-specific estimates can be calculated using data for F and P collected at individual facilities. The particle guidelines in Tables A-1 and A-2 represent the maximum concentrations of particles that can be present in effluents without exceeding the treatment objectives for occluded microorganisms. Table 6-4 contains a summary of particle guidelines containing pathogens in the plant effluent, based on exposure to pathogens, for three different exposure routes. Similar to the concentrations calculated based on indicator organisms, the guidelines were calculated based on the assumption that both F and P are equal to one.

**Table 6-4. Summary of Particle Guidelines Based on Pathogens.[1]**

| Microorganism | Exposure route | Particles containing pathogens/100 mL |
|---|---|---|
| Rotavirus | Drinking water | $5.0 \times 10^{-8}$ |
| Viruses | Reclaimed water | $5.0 \times 10^{-9}$ |
| Rotavirus | Recreational fresh water | $1.0 \times 10^{-2}$ |
| Rotavirus | Recreational marine water | $2.6 \times 10^{-2}$ |

[1]Particle guidelines were calculated using Equation 2-1 and are based on the assumption that F=1 and P=1.

The particle guidelines shown in Table 6-4 represent the maximum allowable particle concentrations (for particles larger than 11 μm) in the effluent to meet the effluent limits for the target microorganisms indicated, assuming that the disinfection system is adequately designed and operated to inactivate all of the dispersed target microorganisms. Particle guidelines are based on target microorganisms and various exposure scenarios. In wastewater there potentially are numerous pathogens present, and risk exists from all pathogens simultaneously present. In this work, scenarios with different pathogens and exposure routes were evaluated and the most conservative guidelines were selected. Particle guidelines are unique to individual treatment facilities, as the potential exposure routes, frequency of occurrence of microorganisms with particles, and microbial density within particles vary among treatment facilities.

### 6.2.5 Particle Guidelines Based on Pathogens, Updated With Measured F Data

The particle guidelines shown in Table 6-4 were developed based on assumptions that the frequency of association and the microbial density are both one (F=1 and P=1). The guidelines were updated to reflect the frequency of association that was measured for total coliform bacteria at each facility (Table 3-2) as shown in Table 6-5. It was assumed that the target pathogens, Rotavirus and enteric viruses, will associate with particles at a frequency that is equivalent to that measured for total coliform bacteria. This assumption appears conservative since total coliform bacteria are typically present in higher concentrations than pathogens in wastewater (Rose et al., 1996; Rose et al., 2001; Rose et al., 2004).

**Table 6-5. Particle Guidelines (# / 100 mL) Based on Pathogens, with Measured Particle Concentrations in Secondary Effluent Samples Shown for Comparison.**[1]

| Pathogen basis | Exposure route | Facility A | Facility B | Facility C | Facility D | Facility E | Facility F | Facility G |
|---|---|---|---|---|---|---|---|---|
| Rotavirus | Drinking water | 1.2E-05 | 1.6E-05 | 3.4E-04 | 5.7E-06 | 2.4E-05 | 8.8E-04 | 3.5E-06 |
| Viruses | Reclaimed water | 1.2E-06 | 1.6E-06 | 3.4E-05 | 5.7E-07 | 2.4E-06 | 8.8E-05 | 3.5E-07 |
| Rotavirus | Recreational fresh water | 2 | 3 | 69 | 1 | 5 | 175 | 1 |
| Rotavirus | Recreational marine water | 6 | 8 | 179 | 3 | 12 | 456 | 2 |
| Measured particle (>11 μm) concentrations (#/100 mL) | | 4900 | 3000 | 160,000 | 81,000 | 82,000 | 63,000 | 150,000 |

[1]Particle guidelines were calculated using Equation 2-1. Measured values of F were used (based on total coliform bacteria) and P was assumed to be one. Particle guidelines represent the effluent limits for particles to meet the risk-based criteria for the indicated pathogens, assuming that the dispersed target pathogens are completely inactivated and that the occluded target pathogens are not inactivated in the disinfection system.

Similar to the particle guidelines based on indicator organisms, the low frequencies of association (F) measured in this study cause the calculated particle guidelines to be higher than if higher values of F had been used. A more conservative approach would be to assume that the frequency of association is one and to use the particle guidelines shown in Table 6-4. A value of one was assumed for P in the absence of available data. Information regarding F and P would facilitate extension of the optimization protocols to a pathogen basis.

The particle guidelines shown in Table 6-5 represent effluent particle concentrations that are not feasible with current treatment and filtration technologies, especially given the potential for the formation of particles in effluent pipelines and channels. The low values in Table 6-4 highlight the importance of designing and operating the disinfection system to sufficiently inactivate pathogens that may be present in effluent particles instead of relying solely on the filtration system. In all cases, the particle concentrations measured for particles larger than 11 μm, which are shown along with the particle guidelines in Table 6-5, are higher than the particle guidelines.

When comparing the particle guidelines in Table 6-5 to those calculated with an indicator organisms basis (Table 5-2), it is clear that if filtration systems are designed to remove particles such that pathogens are sufficiently removed then the guidelines with an indicator organisms basis will be met. Conversely, meeting the particle guidelines for indicator organisms will not guarantee that the removal of occluded pathogens is sufficient. The particle guidelines shown in Table 6-5 are based on meeting the limits for occluded pathogens in particles and do not address inactivation of dispersed microorganisms. Appropriate chlorine doses must be selected and applied to inactivate dispersed microorganisms as well as the occluded pathogens.

## 6.3 Optimization Protocols Based on Pathogens

### 6.3.1 Inactivation Rate Coefficients for Pathogens

Although the radial pore diffusion model was calibrated using total coliform data, the estimates of intra-particle chlorine concentrations can be used to predict the fate of theoretically occluded pathogens. Inactivation coefficients for pathogens are needed to predict the fate of the potentially occluded pathogens. Data reported in published disinfection studies were used to calculate first-order inactivation rate coefficients for pathogens using Equation 1-2 (Table 6-6). Some manipulation of the data reported by the authors listed in Table 6-6 was required to obtain inactivation rate coefficients with the same units as the ones used in this study. For example, in the study by Rice et al. (1999) a linear regression was performed to identify the inactivation rate coefficient. Coincidentally, in the study by Rice et al. (1999) the inactivation rate coefficient for wild type *E. coli* was almost identical (0.86 L/mg/min) to that observed for *E. coli* O157:H7 isolates from cattle (0.89 L/mg/min). In many cases, CT values reported in the referenced studies were converted to inactivation rate coefficients in adherence with Equation 1-2. The data in Table 6-6 are intended to present a range of inactivation rate coefficients for pathogens and is not intended to be an exhaustive review. For additional disinfection information for a variety of microorganisms, refer to Sobsey (1989).

**Table 6-6. Inactivation Rate Coefficients for Pathogens.**

| Microorganism | Type of Disinfectant[1] | Temp. (°C) | pH | $k_i$ (L/mg*min) | Reference |
|---|---|---|---|---|---|
| *Cryptosporidium* oocysts | FC | 22 | 6/8 | 0.00081 | Gyurek et al., 1997 |
| *Cryptosporidium* oocysts | MC | 22 | 8 | 0.0018 | Gyurek et al., 1997 |
| *Bacillus anthracis* spores | MC | 5-25 | 8 | 0.0028 | Rose et al., 2007 |
| Hepatitis A | MC | 5 | 8 | 0.0079 | Sobsey et al., 1988 |
| Coxsackievirus B5 | MC | 5 | 8 | 0.0089 | Sobsey et al., 1988 |
| *Brucella suis*, *Brucella metiensis* | MC | 5-25 | 8 | 0.048 | Rose et al., 2007 |
| *Burkholderia pseudomallei*, *Burkholderia mallei* | MC | 5-25 | 8 | 0.071 | Rose et al., 2007 |
| *Francisella tularensis* | MC | 5-25 | 8 | 0.11 | Rose et al., 2007 |
| *Yersinia pestis* | MC | 5-25 | 8 | 0.13 | Rose et al., 2007 |
| *Encephalitozoan intestinalis* spores | FC | 5 | 6-7 | 0.14 | John et al., 2005 |
| Poliovirus | FC | 5 | 6 | 0.4 | Shin and Sobsey, 2008 |
| *Helicobacter pylori* | MC | 15 | 7 | 0.48 | Baker et al., 2002 |
| Coxsackievirus B5 | FC | 5 | 7 | 0.77 | Sobsey et al., 1988 |
| *E. coli* O157:H7 | FC | 5 | 7 | 0.89 | Rice et al., 1999 |
| Norovirus | FC | 5 | 6 | 4.3 | Shin and Sobsey, 2008 |
| Hepatitis A | FC | 5 | 7 | 5.1 | Sobsey et al., 1988 |
| *Helicobacter pylori* | FC | 15 | 7 | 15 | Baker et al., 2002 |
| Avian influenza virus (HPAI) subtype H5N1 | FC | 5 | 7/8 | 16 | Rice et al., 2007 |

[1]Free chorine=FC, monochloramine=MC

Of the inactivation rate coefficients for pathogen shown in Table 6-6, the inactivation rate coefficient for *Cryptosporidium* and *Bacillus* spores are so low that very little inactivation is expected when chlorine disinfection is used. The inactivation rates for disinfection of Hepatitis A and *Coxsackievirus* B5 with monochloramine obtained using the data of Sobsey et al. (1988) were also very low. The inactivation rate coefficients for many of the pathogens listed in Table 6-6 are within the range of values measured for total coliform bacteria during this study, which was 0.094 to 14 (Table 3-3). To evaluate the inactivation of pathogens potentially occluded in particles two variables for occluded pathogens were added to the model, which will be referred to as Model Pathogen 1 and Model Pathogen 2. The inactivation rate coefficient that was used for Model Pathogen 1 was 0.05 L/mg*min, which was intended to represent the lower range of values shown in Table 6-6 without using a value that was so low that essentially no inactivation would be observed. The inactivation rate coefficient that was used for Model Pathogen 2 was 0.50 L/mg*min, which was intended to represent some of the middle-range values shown in Table 6-6.

### 6.3.2 Estimated Concentration of Occluded Pathogens Following Disinfection

Equation 1-2, in combination with the model output and the published inactivation rate coefficients, was used to predict the concentration of occluded pathogens that would survive a 90-minute disinfection period. Estimates of Model Pathogen 1 presence in effluent particles are shown in Table 6-7 and the estimates for Model Pathogen 2 are shown in Table 6-8. In all cases the viruses were conservatively assumed to be present in the center of particles and the initial concentrations were assumed to be equivalent to that of the total coliform bacteria that were observed in the disinfection tests. Pathogens were conservatively assumed to be present in the largest particles; the same assumption was made for total coliform bacteria.

In Tables 6-7 and 6-8 the shaded numbers represent simulations where the calculated pathogen concentrations were lower than $5x10^{-9}$ PFU/100 mL, the lowest concentration of pathogens that formed the basis for the particle guidelines shown in Table 6-4. The shaded numbers represent scenarios where the combination of particle counts and initial chlorine concentration resulted in acceptable treatment, as defined here by the ability to inactivate occluded pathogens. The guideline of $5x10^{-9}$ PFU/100 mL was not met in any of the estimated Model Pathogen 1 concentrations as shown in Table 6-7. However, the guideline was met for many of the treatment scenarios for the partially-nitrified and non-nitrified effluents as shown Table 6-8. Simulations were completed with progressively higher chlorine concentrations to identify treatment combinations that resulted in inactivation of Model Pathogen 2, the pathogen with the higher inactivation rate coefficient. Simulations with initial chlorine concentrations up to approximately 100 mg/L were completed for Facilities A and B, but none of the simulations yielded occluded pathogen concentrations that were below the treatment objective.

**Table 6-7. Occluded Model Pathogen 1 (PFU/100 mL) Surviving 90-minutes, Calculated Using Model Output.**[1]

| | Initial chlorine conc. (mg/L)[2] | Alterations to the PSD | | | |
|---|---|---|---|---|---|
| | | No change | 30% removal | 60% removal | 90% removal |
| **Facility A** | **70** | 2E+01 | 1E+01 | 7E+00 | 2E+00 |
| | **81** | 2E+01 | 1E+01 | 7E+00 | 2E+00 |
| | **92** | 2E+01 | 1E+01 | 7E+00 | 2E+00 |
| | **103** | 2E+01 | 1E+01 | 7E+00 | 2E+00 |
| **Facility B** | **68** | 1E+01 | 8E+00 | 5E+00 | 1E+00 |
| | **78** | 9E+00 | 7E+00 | 5E+00 | 8E-01 |
| | **88** | 8E+00 | 6E+00 | 4E+00 | 7E-01 |
| | **98** | 8E+00 | 6E+00 | 4E+00 | 6E-01 |
| **Facility C** | **28** | 3E-01 | 3E-01 | 2E-01 | 5E-02 |
| | **38** | 6E-02 | 6E-02 | 4E-02 | 1E-02 |
| | **48** | 1E-02 | 1E-02 | 8E-03 | 2E-03 |
| | **58** | 2E-03 | 2E-03 | 2E-03 | 4E-04 |
| **Facility D** | **66** | 1E+00 | 9E-01 | 5E-01 | 1E-01 |
| | **76** | 8E-01 | 6E-01 | 3E-01 | 6E-02 |
| | **86** | 6E-01 | 4E-01 | 2E-01 | 4E-02 |
| | **96** | 4E-01 | 3E-01 | 1E-01 | 3E-02 |
| **Facility E** | **33.6** | 3E+00 | 4E-01 | 2E-01 | 1E-01 |
| | **43.6** | 2E+00 | 2E-01 | 1E-01 | 6E-02 |
| | **53.6** | 8E-01 | 1E-01 | 6E-02 | 3E-02 |
| | **63.6** | 4E-01 | 6E-02 | 3E-02 | 2E-02 |
| **Facility F** | **24** | 0E+00 | 0E+00 | 0E+00 | 0E+00 |
| | **34** | 9E-01 | 9E-01 | 5E-01 | 1E-01 |
| | **44** | 4E-01 | 3E-01 | 2E-01 | 5E-02 |
| | **54** | 1E-01 | 1E-01 | 7E-02 | 2E-02 |
| **Facility G** | **21.5** | 4E-01 | 3E-01 | 2E-01 | 4E-02 |
| | **31.5** | 2E-02 | 1E-02 | 8E-03 | 2E-03 |
| | **41.5** | 1E-03 | 8E-04 | 4E-04 | 1E-04 |
| | **51.5** | 6E-05 | 4E-05 | 2E-05 | 5E-06 |

[1]Shaded numbers represent concentrations where the virus limit of 5E-9 PFU/100 mL in Table 6-6 is met.
[2]Initial chlorine concentration is the initial model concentration plus an instantaneous chlorine demand.

Table 6-8. Occluded Model Pathogen 2 (PFU/100 mL) Surviving 90-minutes, Calculated Using Model Output.[1,2]

| | Initial chlorine conc. (mg/L)[2] | Alterations to the PSD | | | |
|---|---|---|---|---|---|
| | | No change | 30% removal | 60% removal | 90% removal |
| Facility A | 70 | 8E+00 | 6E+00 | 3E+00 | 8E-01 |
| | 81 | 7E+00 | 5E+00 | 3E+00 | 7E-01 |
| | 92 | 6E+00 | 4E+00 | 3E+00 | 6E-01 |
| | 103 | 5E+00 | 4E+00 | 2E+00 | 5E-01 |
| Facility B | 68 | 2E-01 | 1E-01 | 7E-02 | 1E-02 |
| | 78 | 8E-02 | 6E-02 | 3E-02 | 6E-03 |
| | 88 | 4E-02 | 3E-02 | 2E-02 | 3E-03 |
| | 98 | 2E-02 | 1E-02 | 9E-03 | 1E-03 |
| Facility C | 12 | 2E-06 | 2E-06 | 1E-06 | 4E-07 |
| | 14 | 9E-08 | 9E-08 | 6E-08 | 2E-08 |
| | 16 | 4E-09 | 4E-09 | 3E-09 | 6E-10 |
| | 18 | 2E-10 | 2E-10 | 1E-10 | 3E-11 |
| Facility D | 46 | 6E-07 | 4E-07 | 2E-07 | 4E-08 |
| | 56 | 1E-08 | 1E-08 | 5E-09 | 1E-09 |
| | 66 | 3E-10 | 2E-10 | 1E-10 | 3E-11 |
| | 76 | 8E-12 | 6E-12 | 3E-12 | 6E-13 |
| Facility E | 21.6 | 2E-05 | 3E-06 | 1E-06 | 7E-07 |
| | 23.6 | 5E-06 | 7E-07 | 3E-07 | 2E-07 |
| | 33.6 | 7E-09 | 1E-09 | 5E-10 | 2E-10 |
| | 43.6 | 1E-11 | 1E-12 | 7E-13 | 3E-13 |
| Facility F | 18 | 2E-06 | 1E-06 | 8E-07 | 2E-07 |
| | 20 | 2E-07 | 2E-07 | 1E-07 | 3E-08 |
| | 22 | 3E-08 | 3E-08 | 2E-08 | 5E-09 |
| | 34 | 3E-13 | 3E-13 | 2E-13 | 5E-14 |
| Facility G | 5.5 | 8E-05 | 5E-05 | 3E-05 | 7E-06 |
| | 7.5 | 2E-07 | 1E-07 | 8E-08 | 2E-08 |
| | 9.5 | 6E-10 | 4E-10 | 2E-10 | 5E-11 |
| | 11.5 | 2E-12 | 1E-12 | 6E-13 | 1E-13 |

[1]Shaded numbers represent concentrations where the virus limit of 5E-9 PFU/100 mL in Table 6-6 is met.
[2]Initial chlorine concentration is the initial model concentration plus an instantaneous chlorine demand.

By interpolating between the values shown in Table 6-8 for the partially-nitrified and non-nitrified effluents, combinations of initial chlorine concentration and particle removal were identified that resulted in Model Pathogen 2 concentrations that were below $5x10^{-9}$ PFU/100 mL. The results of the interpolation are shown in Figure 6-1. As shown, when the particle concentrations are reduced, the chlorine required to inactivate the remaining occluded Model Pathogen 2 is reduced. The slopes of the best-fit lines shown in Figure 6-1 indicate the effectiveness of filtration. Where the lines are steeper, more particles must be removed to reduce chlorine requirements, relative to facilities that have best-fit lines with more gentle slopes. The model output for Facilities A and B could not be included in Figure 6-1 because combinations of initial chlorine concentration and particle removal could not be identified that resulted in acceptable treatment.

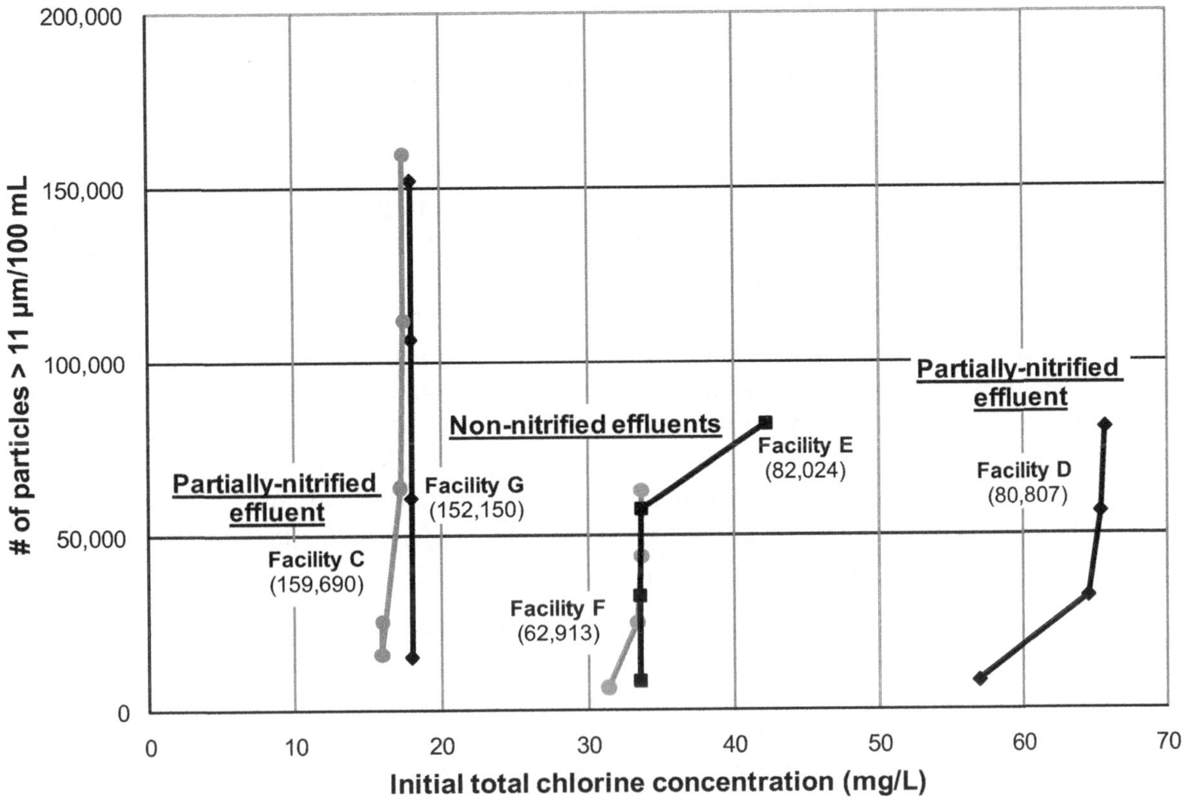


**Figure 6-1. Particle and Initial Total Chlorine Concentrations That Result in Occluded Hepatitis A Concentrations of $5x10^{-9}$ PFU/100 mL (Original Particle Counts Are Listed in Parentheses).**

The for the data from Facilities C and G shown in Figure 6-1 are nearly vertical, indicating that changes in particle concentrations did not influence the concentration of chlorine that was necessary to achieve the treatment objective. However, the concentrations of chlorine required for the Facility C and G effluents are relatively low when compared with requirements for the other effluents: approximately 10-15 mg/L. The data for Facility F are also fairly vertical. The data for Facilities D and E, however, do not have a vertical trend that indicates insensitivity to changes in particle concentrations. In addition, the data from Facilities D and E shown in Figure 6-1 are not linear; the non-linearity might result from the discrete particle sizes that are considered in the model.

## 6.4 Discussion and Conclusions

Although it is premature to establish particle guidelines and optimization protocols that are based on pathogens, a procedure is presented to do so once the necessary data are available. Using the existing regulatory and risk assessment frameworks, calculation of particle guidelines was completed assuming that the frequency of association (F) and density of pathogens in particles containing these organisms (P) were both one. Refer to Chapter 5.0 for a more detailed discussion of the assumptions regarding the variables F and P. The particle guidelines were calculated using the F values measured for the individual facilities; however, use of F=1 yields more conservative particle guidelines and is recommended. In both cases the resulting particles guidelines were low, representing values that are unattainable in current filtration systems, especially given the potential for re-formation of particles in effluent pipelines and channels. The low particular guideline values highlight the importance of reliance on the disinfection system for inactivation of target pathogens. In particular, disinfection systems must be sufficiently designed to inactivated occluded target microorganisms, even those potentially located at the center of effluent particles.

As discussed in Chapter 4.0, the model was first calibrated using total coliform data collected in laboratory studies. In this chapter, the model-generated intra-particle chlorine concentrations, obtained during the simulations, were used to in combination with published inactivation rate coefficients to estimate the concentration of pathogens occluded in effluent particles. The advantage of the optimization protocols is that intra-particle chlorine concentrations provide a means of translating information regarding indicator organisms to estimates of the survival of pathogens potentially present in particles. Several assumptions were made to calculate the estimated concentration of pathogens occluded in particles. The initial concentration of occluded pathogens was assumed to be equivalent to the initial concentration of occluded total coliform bacteria. Occluded pathogens were conservatively assumed to be occluded in the largest particles and were assumed to be located at the center of particles. All calculations were performed assuming a 90-minute contact time.

To extend the current approach to pathogens, appropriate inactivation rate coefficients must be selected. Disinfection data for pathogens are available in published studies, although the data are limited and many of the existing studies were performed using animal strains of pathogens rather than strains pathogenic to humans. Many of the existing disinfection studies were performed using cultured strains in synthetic media, and there may be different results than what would be expected had the experiments been performed using environmental water samples. To represent the range of inactivation rate coefficients gleaned from existing studies, two inactivation rate coefficients were selected. In lieu of using published values of inactivation rate coefficients for pathogens, it would be possible to use a sensitivity analysis for the inactivation rate coefficient variable. First, a range of potential inactivation rate coefficients would be identified based on published values. Then the calculations for occluded pathogen presence could be repeated several times to evaluate the effect of different inactivation rate coefficients.

The effect of initial chlorine concentration and particle removal rates on the inactivation of model pathogens occluded in particles is shown in Tables 6-7 and 6-8. For each facility, managers can identify appropriate chlorine concentrations for a given particle removal rate. Conversely, target particle removal rates can be identified if the chlorine concentration is constrained by facility conditions or chemical costs. Initial chlorine concentrations and particle removal rates were not identified that resulted in acceptable treatment for Model Pathogen 1. Obviously, if simulations using higher chlorine concentrations or higher removal rates were used, the treatment object of $5x10^{-9}$ PFU/100 mL would have been met. However, initial chlorine concentrations and particle removal rates outside of the range of values used in the simulations were deemed infeasible. There are limits regarding operation of the disinfection system that are related to cost and feasibility (e.g. size of the system), and there are limits to the effectiveness of filtration to remove particles (even highly effective filtration systems).

For Model Pathogen 2, that had a higher inactivation rate coefficient than Model Pathogen 1, it was possible to meet the treatment objective for the simulation conditions used for the partially-nitrified and non-nitrified effluents but appropriate treatment conditions for the nitrified effluents were not identified. A preliminary conclusion of the inability of the nitrified effluents to sufficiently reduce occluded pathogens is that the higher decay rates ($k=k'$) and higher partitioning coefficients ($K_d$) result in less available reactive chlorine inside particles (Chapter 4.0 contains a summary of the model parameters selected during calibration).

The information shown in Figure 6-1 is useful for managers of treatment facilities because the effects of operational changes (particle reduction through filtration or changes in chlorine applied) can be evaluated. For example, if a proposed filtration system is capable of reducing the total particle concentration by 50%, the manager can estimate the resulting decrease in total chlorine needed to inactivate occluded pathogens. Of course, disinfection system operating conditions will also need to be selected such that the dispersed target microorganisms are inactivated as well as the occluded ones. Another caveat is that the information shown in Figure 6-1 was developed using a single inactivation rate coefficient that is representative of a range of published values. Appropriate treatment conditions that would result in acceptable concentrations of occluded pathogens for nitrified effluent were not identified. All of the information in Figure 6-1 was obtained assuming that the contact time in the chlorine reactor was 90-minutes. Similar plots could be developed with a different contact time. In general, the results shown in Figure 6-1 indicate the site-specific nature of the impact of filtration (particle removal) on disinfection performance.

A result of the model simulation effort is that disinfection of the nitrified effluents did not result in appreciable reductions in pathogens in the range of treatment conditions that were considered. This conclusion is consistent with the results obtained in Chapter 4.0. Basically, the higher decay rate of free chlorine and the reactivity of free chlorine with organic matter in particles results in lower inactivation rates for microorganisms occluded in particles. The less reactive and more persistent combined chlorine is more effective for reducing concentrations of occluded microorganisms. It should be noted that the conditions considered for the nitrified effluents in the model simulations were considered to be reasonable estimates of initial chlorine concentration (concentrations up to approximately 100 mg/L were considered) and particle removal rates (removal rates up to 90% were considered).

The preliminary results shown in this chapter are based on several assumptions, including the assumption that pathogens are occluded in particles at the same rate as total coliforms. Insufficient data are available to determine the frequency and density of pathogens occluded in effluent wastewater particles (variables F and P). If the initial concentrations of occluded pathogens were assumed lower than the initial concentrations of occluded total coliforms, all of the results shown in this chapter would be lower. Clearly, quantifying the presence of occluded pathogens in effluent wastewater particles would be valuable and would allow for a more meaningful application of the tools presented in this study. The results shown, however, provide preliminary estimates of the pathogen presence in disinfected particles that are conservative because the estimates are based on conservative assumptions.

Extending the current optimization protocols to pathogens would be beneficial because pathogens are the microorganisms of concern in treated effluents. Current data are insufficient to apply the particle guidelines to pathogens and to calculate the estimated survival of pathogens occluded in particles. Included herein is a methodology for applying the optimization approach to pathogens so that the work can be completed in the future when the data are available. The following additional information is needed to extend the optimization protocols to pathogens:

- Frequency of association of pathogens with particles (F),
- Pathogen density within particles that contain pathogens (P), and
- Inactivation rate coefficients for a wider range of pathogens using free and combined chlorine as the disinfecting agent.

# CHAPTER 7.0

# CONCLUSIONS AND RECOMMENDATIONS

The optimization protocols were applied at seven facilities, representing a range of treatment trains and operating conditions. The results indicate that there are a vast number of particles in post-filtration effluents, and that these particles impact the disinfection system performance. In general, as was evident by the inability of most of the facilities to meet the particle guidelines, filtration systems cannot reliably remove a sufficient quantity of particles containing microorganisms; disinfection systems must be designed and operated such that chemical disinfectants saturate the particles and inactivate the occluded microorganisms. Based on available published data regarding the inactivation of pathogens, it appears likely that designing disinfection systems to ensure intra-particle chlorine concentrations are sufficient to inactivate occluded total coliform bacteria may not be adequate to inactivate pathogens.

Data collected in laboratory tests and information collected using the model indicate that there are trends between operating conditions (e.g. MCRT) and optimization of filtration and disinfection systems. However, the data also highlight the site-specific nature of disinfection, suggesting that investigation at each facility is ideal. Multiple analyses conducted at a single facility revealed temporal changes in particle and disinfection characteristics; however, the data were consistent and defined a range of values at a single facility. Additional testing during different seasons and operating conditions is recommended to better understand the range of disinfection and particle characteristics. Managers of treatment facilities have some discretion regarding adjustments of their filtration or disinfection systems, and the results of this study provide guidance in operating these systems to collectively reduce the presence of occluded microorganisms in treated effluents.

## 7.1 Laboratory Data Collected From Seven Facilities

The optimization approach was applied to samples collected from seven facilities, each of which uses a different treatment train and has different operating conditions. For example, the MCRT values amongst the facilities varied from approximately two days to 40 days; consequently, the ammonia nitrogen concentrations measured in this study also varied. In general, the nitrified secondary effluent samples had lower measured turbidity, lower COD values, and lower particle counts. Data collected for the partially-nitrified (ammonia nitrogen concentration of approximately 1 mg/L) and non-nitrified (ammonia nitrogen concentration greater than 20 mg/L) samples reflected higher measured turbidity, higher COD values, and higher particle counts, although the water quality parameter were not directly correlated to MCRT or the presence of ammonia. All samples contained thousands of large particles (with diameters larger than 11 μm) per 100 mL; the highest concentration of particles observed was 160,000 particles with diameters larger than 11 μm per 100 mL. Although higher turbidity measurements generally correlated with higher particle counts, the relationship was not strong, suggesting that PSD is a better indicator of particle presence.

Disinfection tests were performed to generate data in the tailing region of the inactivation curves, which were essential for calibration of the model. Variable amounts of chlorine were required in disinfection tests to produce inactivation curves with identifiable tailing regions and measurable concentrations of target microorganisms. Selection of appropriate doses was critical, as addition of too little chlorine did not produce tailing and too much chlorine resulted in concentrations of target microorganisms that were below detection limits. For optimal model calibration it is preferable to select an initial chlorine concentration that produces as much tailing as possible, but it is difficult to predict the most appropriate initial chlorine concentration. Completion of disinfection tests was further complicated by the low concentrations of target microorganisms detected in some of the samples in this study. This study was conducted using total coliform bacteria as the target microorganism; however, there is some freedom over selection of a target microorganism although the ideal candidate would be abundantly present and not exhibiting significant tailing phenomena other than occlusion in particles during disinfection.

PSD data collected prior to and following filtration revealed the effectiveness of filtration; however, substantial concentrations of particles were observed in post-filtration effluents. Particle removal rates were higher for larger particles and lower for smaller particles. For most particle sizes, removal rates were greater than 50% and often near 100% for the largest particles. The post-filtration PSD data verified that it is reasonable to assume that a fraction of particles with diameters greater than 11 μm are removed in filtration. The data also suggest that large particles that can potentially impact disinfection performance are commonly found in post-filtration effluents and that the disinfection systems should be designed and operated to address these particles.

The optimization protocols were completed three additional times at Facility A to perform a model simulation exercise and further investigate the approach of using the radial pore diffusion model for disinfection of occluded microorganisms. The experimental data collected from Facility A in sequential experiments reflected some temporal variability, although measured water quality values and disinfection test result were consistent for Facility A as compared with results from other facilities. Additional testing at a single facility, such as Facility A, would be beneficial to further clarify the temporal variability of disinfection and particle characteristics. Ideally, tests would be conducted to reflect diurnal and seasonal effects as well as the impact of process upsets.

Two changes could be incorporated into the optimization protocols to streamline the procedures and facilitate implementation. Measurement of particle wet density was completed for each of the facilities to ascertain the particle porosity and volumetric void ratio, a parameter that is necessary to run the model. The particle porosity was not highly variable, suggesting that the particle wet density test could be eliminated from the optimization protocols or at least completed only once or twice at each facility. Incorporation of automated particle counters would enhance application of the optimization protocols. The particle counting method used in this study is time-consuming and labor-intensive. An easier and faster method for collecting PSD data would streamline optimization efforts.

## 7.2 Application of a Radial Pore Diffusion Model for the Seven Facilities

A radial pore diffusion model was used that considered mass transfer of chlorine from the bulk fluid of a reactor into spherical particles of four distinct sizes. The model was calibrated using data collected in the laboratory studies and simulations were run with varying initial chlorine concentrations and particle counts. Specifically, the constituents needed to run the model are PSD, initial chlorine concentration, and the particle volumetric void ratio. Chlorine residual data that were measured during the disinfection test were used to calibrate the model as were concentrations of total coliform bacteria in the tailing region of the disinfection tests.

Parameter estimating software (UCODE) was used to select parameters that provided a best-fit of the model-calculated estimates with the observed data. Fitting parameters consisted of the chlorine decay rates ($k=k'$) and a partitioning coefficient ($K_d$). In general, the model calculations for chlorine concentrations in the bulk fluid provided a good fit with the experimentally measured data although the model fits for occluded total coliform were not as good. The model estimates for occluded total coliform bacteria could likely be improved by changing the assumptions regarding occlusion of total coliform in particles and the locations of the total coliform bacteria within the particles. To change the assumptions, the model calibration would need to be repeated. Currently the total coliform curves produced by the model contain lag regions that are the result of diffusion limitations in the various size particles. By assuming that total coliform bacteria occupy a range of particle sizes and that they are located throughout particles, the curves could become more linear. The current assumption that occluded total coliform bacteria are located in the largest particles and that they are located at the particle center is conservative and was intended to ensure that all particles are sufficiently saturated with chlorine to sufficiently inactivate the occluded total coliform bacteria.

The facility characteristics and measured water quality data correlated reasonably with the selected model parameters for the calibrated models. The nitrified effluents had higher decay rates and partitioning coefficients that indicated free chlorine was more reactive with the organic matter inside of particles. In this study the Biot number was assumed consistent for all the facilities and all of the particle sizes. An improvement could be made to the model by locating appropriate values of $k_m$, the mass transfer coefficient across the particle surface, and calculating values for the Biot number. An attempt was made to identify values of $k_m$, however, the resulting Biot numbers were too high and the mass transfer into the particles was excessive and a good fit for the bulk fluid chlorine observations was not obtained. It is likely that the methods used to develop the published value of $k_m$ were not appropriate for this study.

The intra-particle chlorine concentrations generated by the model reflect the dynamic nature of chlorine disinfection, where the concentration of chlorine in the bulk fluid is constantly decreasing and the flux of chlorine into particles is constantly changing as a result. As a result of boundary conditions at the particle surface and chlorine decay kinetics, a stable condition where the intra-particle chlorine concentrations are equivalent to the concentrations in the bulk solution is never obtained. There is no steady-state condition and estimates of microbial inactivation in particles is best estimated using numerical models that update intra-particle chlorine concentrations and microbial concentrations in sequential time steps.

## 7.3 Optimization Protocols for the Seven Facilities

Model simulations were run for various initial chlorine concentrations and the following particle removal conditions: 0% removal, 30% removal, 60% removal, and 90% removal. The outcomes of the model simulations were estimates of bulk fluid chlorine, intra-particle chlorine, occluded total coliform, and occluded pathogen concentrations. Two scenarios were used for the pathogens, consisting of two different inactivation rate coefficients that were intended to represent the range of values reported in the literature. A 90-minute contact time was selected in all cases and the inactivation rate coefficients observed in the laboratory studies were used. All microorganisms were conservatively assumed to be occluded in the largest particles and pathogens were assumed to be present in particles at the same rate as total coliform bacteria.

In addition to estimating the survival of occluded total coliform bacteria, the particle guidelines were updated using measured values of the frequency of association (F). The resulting particle guidelines are not as conservative as those based on the assumption that F is one because experiments to determine F yielded values that were lower than expected, possibly due to the low dose of UV light that was used to irradiate dispersed microorganisms and may have inadvertently irradiated a portion of the occluded fraction. The tests for microbial density inside of particles containing microorganisms (P) yielded insight into the density of heterotrophic bacteria and aerobic endospores in particles; however, it is difficult to ascertain accurate estimates of the microbial density inside of particles given the stochastic nature of P. Hence, the particle guidelines, based on indicator organisms and assuming F and P values of one, are recommended. The most conservative particle guidelines were based on the detection limits of the assays for total coliform, fecal coliform, and *E. coli.*, which is a microbial limit for total coliform of 2 MPN/100 mL was adopted. Although the observed particle counts were lower than the particle guidelines for some of the facilities included in this study, the particle guidelines were not met in every case, indicating that the disinfection systems must be designed and operated to ensure inactivation of occluded total coliform bacteria.

Calculations of occluded total coliform bacteria were compared with the microbial limit of 2 MPN/100 mL to determine combinations of particle and initial chlorine concentrations that resulted in acceptable treatment. In particular, the estimates revealed the sensitivity of chlorine disinfection system performance to changes in particle counts achieved using filtration. For some of the facilities in the study, large decreases in particle counts appear necessary to significantly reduce the chlorine required. The results are useful for managers of treatment facilities for making decisions regarding their filtration and disinfection systems. As an example, managers can identify targets for particle removal given constraints regarding chlorine dose (both initial chlorine concentration and contact time). Similarly, the particle removal percentage can be constrained to identify the resulting minimum initial chlorine concentration.

For the facilities included in this study, the nitrified effluents had lower concentrations of particles, and a modest reduction in particle concentration (e.g. 60% removal) resulted in significant decreases in the chlorine required to meet the microbial discharge limit. However, it may be difficult to remove the particles in the nitrified secondary effluents because the initial concentrations of particles are low. Results from the facilities with non-nitrified and partially-nitrified effluents had mixed results, indicating the site-specific nature of disinfection and the effects of particles. In all cases, however, removal of particles had the effect of lowering the amount of chlorine required to inactivate the occluded microorganisms.

All of the numbers presented in Chapters 5.0 and 6.0 were calculated based on a 90-minute chlorine contact time; however, the optimization approach could be repeated using a different contact time by using model estimates of intra-particle chlorine concentrations for another time step. In the model simulations, constant particle removal was assumed for all particle size classes, but it is possible to make different assumptions regarding particle removal. For example, a simulation could be run where all particles larger than a threshold value were removed.

The experimental and modeling work completed in this study support the theory that diffusion limitations impact the inactivation of microorganisms occluded in effluent wastewater particles. Higher concentrations of chlorine and removal of particles reduced the degree of tailing observed in disinfection tests. The radial pore diffusion model was calibrated to describe the mass transfer of chlorine from the bulk aqueous phase into the particles. In general, decay and intra-particle sorption was more significant when disinfecting the nitrified effluents, suggesting that free chlorine is less stable and more reactive than combined chlorine products. A similar phenomenon was observed by LeChevallier et al. (1988) in a study where combined chlorine was more effective than free chlorine for inactivating bacteria enmeshed in biofilms, which is somewhat surprising because free chlorine is more effective for inactivating dispersed microorganisms (Olivieri, 1985).

## 7.4 Inherent Assumptions and Limitations of the Optimization Approach

Although the optimization protocols are a useful tool for the treatment industry, the assumptions and limitations of the current procedures must be understood when applying the approach. One assumption is that occlusion of the target microorganisms in particles is the predominant tailing mechanism in disinfection systems. A target microorganism must be selected that does not exhibit other significant tailing mechanisms in disinfection tests. For example, heterotrophic bacteria are less useful as model organisms in optimization tests because there is a sub-population of heterotrophic bacteria that are extremely resistant to chlorine and persist even when high doses of chlorine are applied. Another important assumption is that occluded and dispersed target microorganisms adhere to first-order inactivation kinetics and have the same inactivation rate coefficients; the only factor limiting the inactivation of occluded microorganisms is the diffusion of chlorine into particles. Verification of the inactivation characteristics of occluded and dispersed target microorganisms would be of interest for validating the approach, but it would be difficult to isolate the occluded fraction of microorganisms without injury to cells or disruption of the population dynamics. Additional assumptions include assuming that microorganisms are preferentially located in larger particles and that they are located throughout particles, including the particle center. It is conservatively assumed that each particle containing microorganisms contains at least one microorganism at the particle center.

To calculate particle guidelines, assumptions must be made regarding the frequency of association of microorganisms in particles (F) and the microbial density of microorganisms in the subset of particles containing microorganisms (P). Data for F is difficult to collect because it is difficult to isolate the occluded microorganisms and ensure that no dispersed microorganisms are present. In this study a low dose of UV light was used to inactivate dispersed microorganisms, and it is likely that the presence of target microorganisms in particles was underestimated because some of the occluded target microorganisms are inactivated. For F the most conservative approach is to assume that it is one because this assumption assures that the stochastic nature of F does not mean that there will be risk of some samples having an F value that is higher than the average value.

Similar to F, values of P are also difficult to ascertain. Experiments were completed to disrupt particles prior to microbial assays was completed in this study to obtain better estimates of P. The results suggest that the density of microorganisms in particles is related to the concentration of the target microorganism in the undisrupted sample. For example, for prevalent microorganisms such as heterotrophic bacteria a high density was obtained. For microorganisms with lower concentrations, such as aerobic endospores, the density appears much lower. A value of one is also suggested for P since this is consistent with current regulations and it is difficult to select a different value given the paucity of available information. To ensure a conservative approach, occluded microorganisms were assumed to inhabit the center of particles, requiring sufficiently high intra-particle chlorine concentrations throughout particles that ensure all occluded microorganisms are inactivated.

Adoption of the radial pore diffusion model also introduced assumptions and uncertainty into the optimization protocols. One assumption inherent in the model is that there are only four particle sizes and that the particles are spherical. An assumption made when using the model is that there is mass transfer resistance across the particle surface. Also, it is assumed that the reactive chlorine products inside of particles behave similarly to chlorine products in the bulk solution, and that a single reactive chlorine species is present. The fate of chlorine products is assumed to adhere to first-order decay kinetics inside particles and in the bulk fluid of the reactor. First-order sorption kinetics were also assumed for reactivity of chlorine species in the particle pore spaces with the organic matter.

## 7.5 Relevance of Health Risks to the Optimization Approach

The current regulatory guidelines for microbial limits in treated effluents are based on health studies, technology limitations, and analytical method limitations. Practices are adopted after a lengthy consultation and study process and are maintained until there is evidence demonstrating that unacceptable health risks are present (e.g. disease outbreaks related to poor water quality). In general, the strength of the existing regulatory framework is that it has historically demonstrated few adverse health risks. In situations where there have been adverse health outcomes related to inadequate regulations, the regulatory community has responded with new regulations.

A major drawback of current regulations is that they are often reliant on monitoring of indicator organisms that are typically more easily removed and inactivated in treatment systems than pathogens. For example, intra-particle chlorine concentrations must be higher to inactivate many target pathogens than commonly used indicator organisms. It is possible that treated effluents are currently meeting discharge limits for indicator organisms, while releasing large quantities of viable pathogens that are protected from disinfectants due to occlusion in effluent particles. Another drawback of current regulations for filtration systems is that they are reliant on monitoring of turbidity to ensure acceptable filtration is occurring. However, turbidity is not easily related to health risks and for highly treated effluents, PSD is a preferred alternative process control parameter because small and large particles impact downstream disinfection systems differently.

While the work presented herein does not necessarily suggest that more stringent regulations should be implemented, it does suggest that treatment processes should be evaluated and monitored to access the particle and disinfection conditions at the individual facilities and ensure that sufficient quantities of occluded microorganisms are removed in filtration systems or are inactivated in disinfection processes. Performance of filtration and disinfection systems is dependent on site-specific conditions and it is difficult to recommend a single treatment metric (e.g. CT concept) that will apply at all facilities.

## 7.6 Adoption of Particle Guidelines

Particle guidelines represent the maximum concentration of particles (with diameters greater than 11 μm) that can be present in an effluent stream, which ensures that the concentration of occluded target microorganisms is below the microbial limit that formed the basis of the particle guideline. Particle guidelines were calculated based on various exposure scenarios, microbial limits, and assumptions regarding the presence of microorganisms in particles. Microbial limits were selected based on existing regulatory guidelines. Particle guidelines only ensure the reduction of occluded target microorganisms; disinfection systems must be in place and effectively operated to inactivation the dispersed of fraction of microorganisms. If a facility cannot meet the particles guidelines, the disinfection system must be designed and operated to ensure that intra-particle disinfectant conditions are such that the occluded target microorganisms are sufficiently inactivated.

Extension of particle guidelines to have a pathogen basis would be ideal. There are existing regulatory guidelines that could form the bases for such particle guidelines. In some cases, a risk assessment framework could be used to calculate discharge limits for various pathogens, which would make it possible to calculate particle guidelines with a pathogen basis. However, additional data are needed regarding the presence of pathogens in particles (variables F and P). The value of F could conservatively be set at one, as was recommended for the particle guidelines with an indicator organism basis. However, assigning F a value of one may be overly conservative given the extremely low concentrations of many pathogens in wastewater. The value of P must be given further consideration as it is not clear which values should be assigned to it and it is not clear that particles containing more than one pathogen present a greater health risk than particles with a single pathogen.

## 7.7 Extension of the Optimization Protocols to a Pathogen Basis

Additional work could potentially reduce the uncertainty of the optimization approach and make it possible to extend the current work from an indicator organism basis to pathogens. Use of the optimization approach to ensure that occluded pathogens in effluent particles are adequately reduced is a worthwhile goal since intra-particle chlorine concentrations that are high enough to inactivate total coliform bacteria may not be high enough to inactivate occluded pathogens.

Estimating survival of pathogens potentially present in wastewater particles was completed using intra-particle chlorine concentrations from the model as well as inactivation rate coefficients for the target pathogens. In this study a contact time of 90-minutes was used and the intra-particle chlorine concentrations corresponding to this contact time were identified. Inactivation rate coefficients were adopted from published disinfection data although the results obtained would be improved if additional data for inactivation of a variety of pathogens using various disinfectants were available. The result of estimating pathogen survival in wastewater particles was that appropriate combinations of initial chlorine and particle concentrations were identified that resulted in estimates of appropriate treatment. The end product was similar to what was achieved in this study for total coliform bacteria except that the results were relevant to removal of pathogens, the microorganisms with public health significance. As shown here, the conditions selected based on total coliform bacteria were not necessarily protective in that pathogens were not sufficiently removed relative to the treatment objective established during the calculation of particle guidelines based on pathogens.

## 7.8 Practicality of the Optimization Protocols

Managers of treatment facilities must decide if resources should be dedicated to improving performance of their filtration or disinfection system to collectively reduce the presence of microorganisms in plant effluents. The basic strategy presented here is to remove particles to the extent possible using filtration and to ensure that the chlorine concentration and contact time are sufficient to ensure that the remaining particles are sufficiently saturated with chlorine to inactivate target microorganisms. For example, it may be possible to enhance particle removal in the filtration system by adding a new coagulation system or upgrading an existing system. There are limits to the potential for optimizing filtration and disinfection systems that are related to capital and operating costs, capability of existing technology, etc. A surprising conclusion of the model simulations for occluded pathogens was that particle removal did not strongly influence the concentration of chlorine necessary to inactivate occluded pathogens. Particle removal was more useful for reducing the amount of chlorine required to inactivate occluded total coliform bacteria. Of course, if very high removal rates of particles were possible (near 100%) the amount of chlorine required would decrease significantly because removal of the microorganisms would be achieved in the filtration process.

In general, managers of treatment facilities should take advantage of existing filtration systems, optimizing and verifying acceptable performance periodically using PSD measurements. Turbidity measurements are not sufficient. Verification should be made that filtration systems are removing a sufficient total number of particles and that they are removing a sufficient fraction of larger particles. Effluent PSD measurements can also be used to verify that re-formation of particles in pipelines and channels that are located downstream of filtration systems is not problematic and that proper maintenance is being performed and is effective. Additionally, PSD measurements can be used to improve performance of existent coagulation systems.

One of the most significant contributions of this research is that the approach provides a meaningful interpretation of PSD data. Unlike turbidity, use of PSD provides more insight on the effectiveness of filtration because particles of different sizes fare differently in disinfection systems. PSD data gives information regarding the potential inactivation of occluded microorganisms. Implementation of automated particle counters would provide managers of treatment facilities greater access to timely data, and such data could be linked with information obtained using the optimization protocols. Log-removal plots (Figure 4-3) can be used in combination with information regarding the PSD to estimate the fate of occluded total coliform bacteria. Plots similar to Figure 4-3 could also be developed for pathogens.

The particle guidelines developed using the pathogen basis are very small, representing numbers that are impossible to achieve using current filtration technology and considering the re-formation of particles in operating treatment facilities. The small numbers suggest that the focus should be on optimizing the disinfection system such that all remaining particles have intra-particle disinfectant concentrations that are high enough to inactivate occluded target microorganisms.

It is difficult to determine if treatment systems are operating optimally, and it is imperative to establish measurable and attainable treatment goals. Typically, a treatment goal is determined by regulations, which impose discharge permit limits, design criteria, operating parameters, etc. How do we decide if a filtration system is working well? The current practice is to establish acceptable filtration rate and turbidity limits for filter effluents. However, the site-specific nature of filtration system performance at individual treatment facilities is not necessarily addressed in current practice. As a result of differences in untreated wastewater quality and performance of the secondary treatment systems, filtration can be more effective at some facilities than at others. For example, the turbidity of filter influent at some facilities is lower than the turbidity of the filter effluent at other facilities. To some extent benchmarking is an effective strategy, as the performance at a single facility is tracked over time and compared with previously collected data.

The optimization protocols are helpful for managers of treatment systems, but they also suggest that several issues should be addressed during design and start-up of a new facility. During design, the conditions and results for similar facilities can be evaluated to better understand the expected range of conditions at the new plant. During start-up, the PSD of secondary and filter effluents should be checked periodically to determine operating conditions that provide PSDs that are most amenable to disinfection. Optimization of coagulations should be performed with PSD verification as well. The combination of chlorine concentrations and PSD should be monitored along with effluent coliform counts, an exercise that can be facilitated using the radial pore diffusion model.

The approach described herein could be used by plant operators and managers to better understand the fate of microorganisms occluded in effluent wastewater particles. However, there are several challenges that would have to be addressed to reliably produce effluents that have sufficiently low concentrations of occluded microorganisms in the treated effluent. Some of the challenges include the following:

- Operational strategies of filtration and disinfection systems that allow for response to plant upsets, seasonal changes, and diurnal changes.
- Appropriate process control loops that incorporate consideration of the removal and inactivation of microorganisms occluded in particles.
- Prompt procurement of quality PSD data.

APPENDIX A

# PARTICLE GUIDELINES

**Table A-1. Risk Assessment Calculations for Particle Guidelines Based on Pathogens.**

| Guide-line | Microorganism | Exposure route | P | α | $N_{50}$ | r | d | Vc (L) | Morbidity ratio | Conc. (org./100 mL) | F | P | Np (#/100 mL) |
|---|---|---|---|---|---|---|---|---|---|---|---|---|---|
| 5 | *Cryptosporidium* | drinking water | 2.7E-7 | | | 0.0042 | 6.5E-5 | 1 | 0.39 | 2.5E-6 | 1 | 1 | 2.5E-6 |
| 5 | *E. coli* ETEC | drinking water | 2.7E-7 | 0.18 | 8.60E+07 | | 2.8E+0 | 1 | | 2.8E-1 | 1 | 1 | 2.8E-1 |
| 5 | *E. coli* O157:H7 | drinking water | 2.7E-7 | 0.21 | 1120 | | 5.6E-5 | 1 | | 5.6E-6 | 1 | 1 | 5.6E-6 |
| 5 | *G. lamblia* | drinking water | 2.7E-7 | | | 0.02 | 1.4E-5 | 1 | 0.5 | 6.9E-7 | 1 | 1 | 6.9E-7 |
| 5 | Rotavirus | drinking water | 2.7E-7 | 0.265 | 5.6 | | 4.6E-7 | 1 | 0.57 | 2.6E-8 | 1 | 1 | 2.6E-8 |
| 5 | *Salmonella* | drinking water | 2.7E-7 | 0.31 | 2.36E+04 | | 2.5E-3 | 1 | 0.22 | 5.5E-5 | 1 | 1 | 5.5E-5 |
| 5 | *Shigella* | drinking water | 2.7E-7 | 0.21 | 1120 | | 5.6E-5 | 1 | | 5.6E-6 | 1 | 1 | 5.6E-6 |
| 6 | *Cryptosporidium* | recreational fresh water | 8.0E-3 | | | 0.0042 | 1.9E+0 | 0.13 | 0.39 | 5.7E-1 | 1 | 1 | 5.7E-1 |
| 6 | *Cryptosporidium* | recreational marine water | 1.9E-2 | | | 0.0042 | 4.6E+0 | 0.13 | 0.39 | 1.4E+0 | 1 | 1 | 1.4E+0 |
| 6 | *E. coli* ETEC | recreational fresh water | 8.0E-3 | 0.18 | 8.6E+7 | | 8.5E+4 | 0.13 | | 6.6E+4 | 1 | 1 | 6.6E+4 |
| 6 | *E. coli* ETEC | recreational marine water | 1.9E-2 | 0.18 | 8.6E+7 | | 2.1E+5 | 0.13 | | 1.6E+5 | 1 | 1 | 1.6E+5 |
| 6 | *E. coli* O157:H7 | recreational fresh water | 8.0E-3 | 0.21 | 1120 | | 1.7E+0 | 0.13 | | 1.3E+0 | 1 | 1 | 1.3E+0 |
| 6 | *E. coli* O157:H7 | recreational marine water | 1.9E-2 | 0.21 | 1120 | | 4.1E+0 | 0.13 | | 3.2E+0 | 1 | 1 | 3.2E+0 |
| 6 | *G. lamblia* | recreational fresh water | 8.0E-3 | | | 0.02 | 4.0E-1 | 0.13 | 0.5 | 1.5E-1 | 1 | 1 | 1.5E-1 |
| 6 | *G. lamblia* | recreational marine water | 1.9E-2 | | | 0.02 | 9.6E-1 | 0.13 | 0.5 | 3.7E-1 | 1 | 1 | 3.7E-1 |
| 6 | Rotavirus | recreational fresh water | 8.0E-3 | 0.265 | 5.6 | | 1.4E-2 | 0.13 | 0.57 | 6.0E-3 | 1 | 1 | 6.0E-3 |
| 6 | Rotavirus | recreational marine water | 1.9E-2 | 0.265 | 5.6 | | 3.3E-2 | 0.13 | 0.57 | 1.5E-2 | 1 | 1 | 1.5E-2 |
| 6 | *Salmonella* | recreational fresh water | 8.0E-3 | 0.31 | 2.4E+4 | | 7.4E+1 | 0.13 | 0.22 | 1.3E+1 | 1 | 1 | 1.3E+1 |
| 6 | *Salmonella* | recreational marine water | 1.9E-2 | 0.31 | 2.4E+4 | | 1.8E+2 | 0.13 | 0.22 | 3.1E+1 | 1 | 1 | 3.1E+1 |

| Guide-line | Microorganism | Exposure route | P | α | $N_{50}$ | r | d | Vc (L) | Morbidity ratio | Conc. (org./100 mL) | F | P | Np (#/100 mL) |
|---|---|---|---|---|---|---|---|---|---|---|---|---|---|
| 7 | *Shigella* | recreational fresh water | 8.0E-3 | 0.21 | 1120 | | 1.7E+0 | 0.13 | | 1.3E+0 | 1 | 1 | 1.3E+0 |
| 7 | *Shigella* | recreational marine water | 1.9E-2 | 0.21 | 1120 | | 4.1E+0 | 0.13 | | 3.2E+0 | 1 | 1 | 3.2E+0 |

[1]See Table 6-2 for an explanation

**Table A-2. Particle Guidelines Based on Indicator Organisms and Pathogens.**

| Guideline[1] | Microorganism | Exposure route | Conc. (org./100 mL) | F | P | Np (#/100 mL) |
|---|---|---|---|---|---|---|
| A 1 | *E. coli* | recreational fresh water | 1.26E+2 | 1 | 1 | 1.26E+2 |
| A 1 | enterococci | recreational fresh water | 3.3E+1 | 1 | 1 | 3.3E+1 |
| A 1 | enterococci | recreational marine water | 3.5E+1 | 1 | 1 | 3.5E+1 |
| A 1 | fecal coliform | recreational fresh and marine water | 2.0E+2 | 1 | 1 | 2.0E+2 |
| A 2 | total coliform | reclaimed water | 2.3E+1 | 1 | 1 | 2.3E+1 |
| A 2, A 3 | total coliform | drinking or reclaimed water | 2.0E+0 | 1 | 1 | 2.0E+0 |
| A 3 | fecal coliform | drinking water | 2.0E+0 | 1 | 1 | 2.0E+0 |
| B 1 | *Cryptosporidium* | reclaimed water | 5.8E-3 | 1 | 1 | 5.8E-3 |
| B 1 | *Cryptosporidium* | reclaimed water | 2.2E-2 | 1 | 1 | 2.2E-2 |
| B 1 | Enterovirus | reclaimed water | 4.4E-5 | 1 | 1 | 4.4E-5 |
| B 1 | Enterovirus | reclaimed water | 1.7E-4 | 1 | 1 | 1.7E-4 |
| B 1 | *Giardia* | reclaimed water | 1.4E-3 | 1 | 1 | 1.4E-3 |
| B 1 | *Giardia* | reclaimed water | 5.0E-3 | 1 | 1 | 5.0E-3 |
| B 3 | *Cryptosporidium* | drinking water | 1.5E-1 | 1 | 1 | 1.5E-1 |
| B 3 | *Giardia* | drinking water | 6.6E-1 | 1 | 1 | 6.6E-1 |
| B 3 | viruses | drinking water | 6.2E-2 | 1 | 1 | 6.2E-2 |
| B 4 | viruses | reclaimed water | 5.0E-9 | 1 | 1 | 5.0E-9 |
| B 5 | *Cryptosporidium* | drinking water | 2.5E-6 | 1 | 1 | 2.5E-6 |
| B 5 | *E. coli* ETEC | drinking water | 2.8E-1 | 1 | 1 | 2.8E-1 |
| B 5 | *E. coli* O157:H7 | drinking water | 5.6E-6 | 1 | 1 | 5.6E-6 |
| B 5 | *G. lamblia* | drinking water | 6.9E-7 | 1 | 1 | 6.9E-7 |
| B 5 | Rotavirus | drinking water | 2.6E-8 | 1 | 1 | 2.6E-8 |
| B 5 | *Salmonella* | drinking water | 5.5E-5 | 1 | 1 | 5.5E-5 |
| B 5 | *Shigella* | drinking water | 5.6E-6 | 1 | 1 | 5.6E-6 |
| B 6 | *Cryptosporidium* | recreational fresh water | 5.7E-1 | 1 | 1 | 5.7E-1 |
| B 6 | *Cryptosporidium* | recreational marine water | 1.4E+0 | 1 | 1 | 1.4E+0 |
| B 6 | *E. coli* ETEC | recreational fresh water | 6.6E+4 | 1 | 1 | 6.6E+4 |
| B 6 | *E. coli* ETEC | recreational marine water | 1.6E+5 | 1 | 1 | 1.6E+5 |
| B 6 | *E. coli* O157:H7 | recreational fresh water | 1.3E+0 | 1 | 1 | 1.3E+0 |
| B 6 | *E. coli* O157:H7 | recreational marine water | 3.2E+0 | 1 | 1 | 3.2E+0 |
| B 6 | *G. lamblia* | recreational fresh water | 1.5E-1 | 1 | 1 | 1.5E-1 |
| B 6 | *G. lamblia* | recreational marine water | 3.7E-1 | 1 | 1 | 3.7E-1 |
| B 6 | Rotavirus | recreational fresh water | 6.0E-3 | 1 | 1 | 6.0E-3 |
| B 6 | Rotavirus | recreational marine water | 1.5E-2 | 1 | 1 | 1.5E-2 |
| B 6 | Salmonella | recreational fresh water | 1.3E+1 | 1 | 1 | 1.3E+1 |
| B 6 | *Salmonella* | recreational marine water | 3.1E+1 | 1 | 1 | 3.1E+1 |
| B 6 | *Shigella* | recreational fresh water | 1.3E+0 | 1 | 1 | 1.3E+0 |
| B 6 | *Shigella* | recreational marine water | 3.2E+0 | 1 | 1 | 3.2E+0 |

[1]'A' refers to guidelines in Table 2-2 and 'B' refers to guidelines in Table 6-2.

APPENDIX B

# PSD DATA

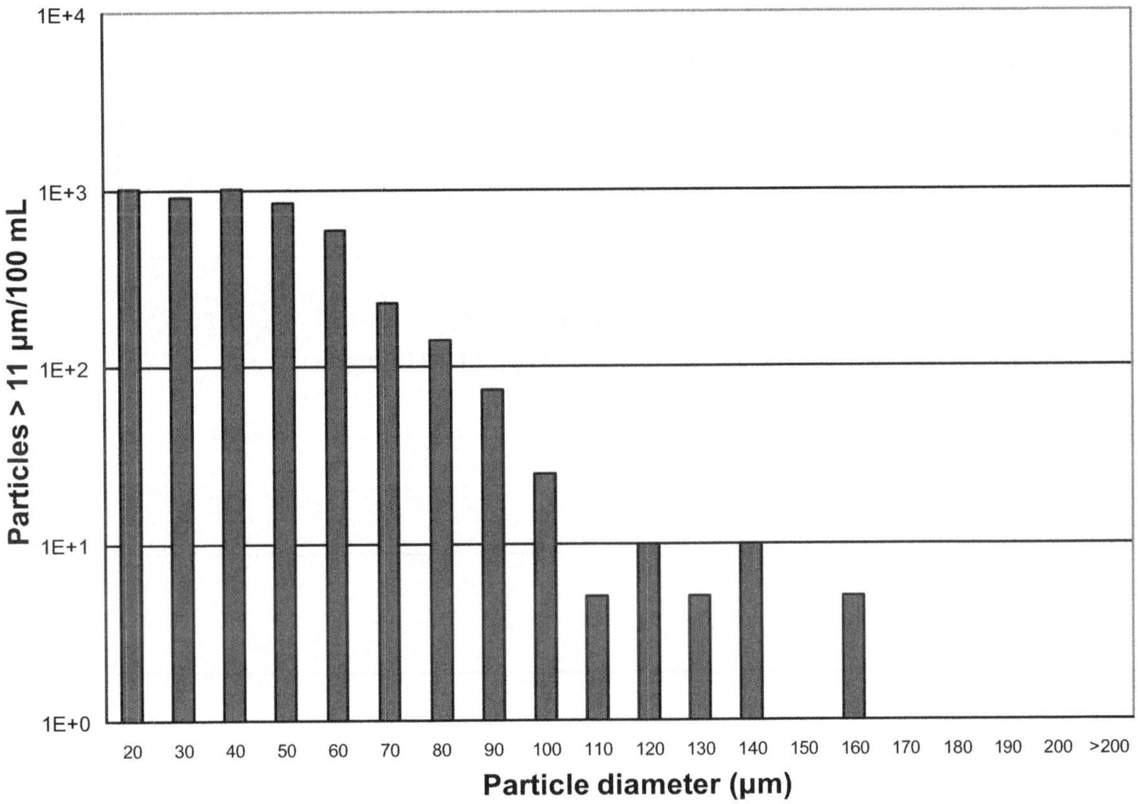


**Figure B-1. Particle Size Distribution, Facility A Secondary Effluent Sample.**

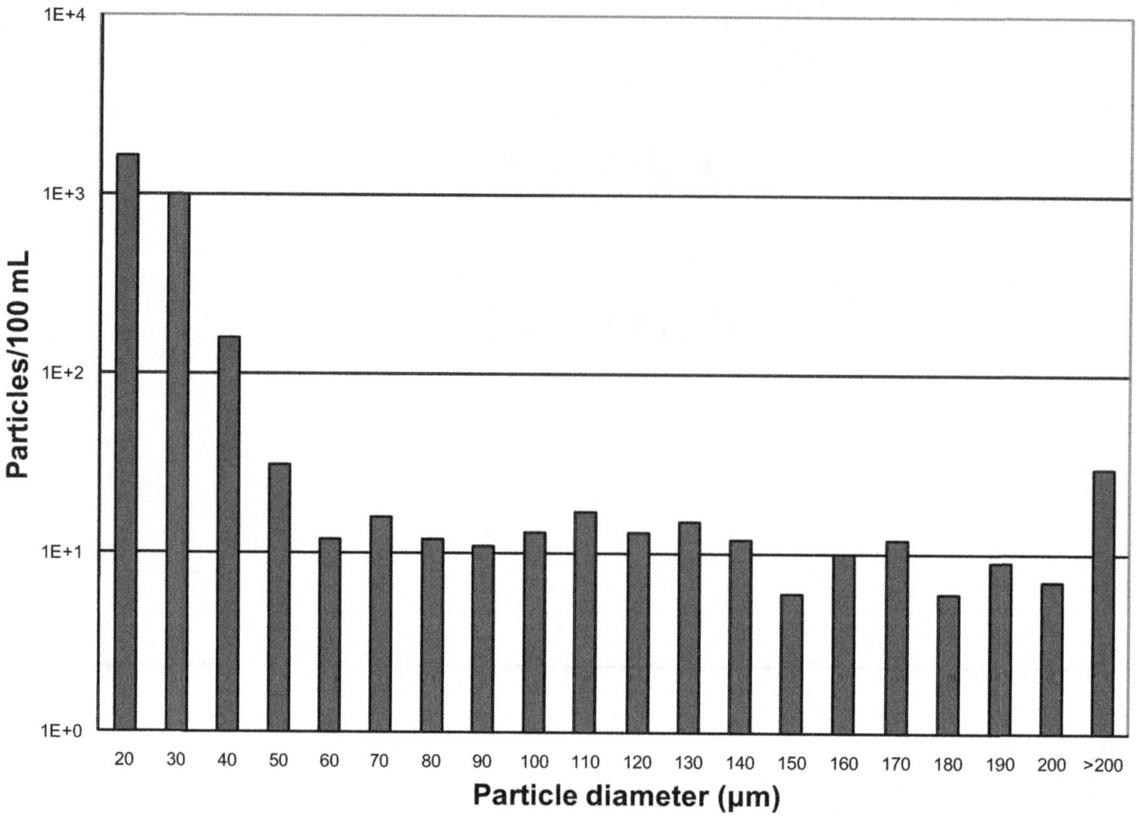


Figure B-2. Particle Particle Size Distribution, Facility B Secondary Effluent Sample.

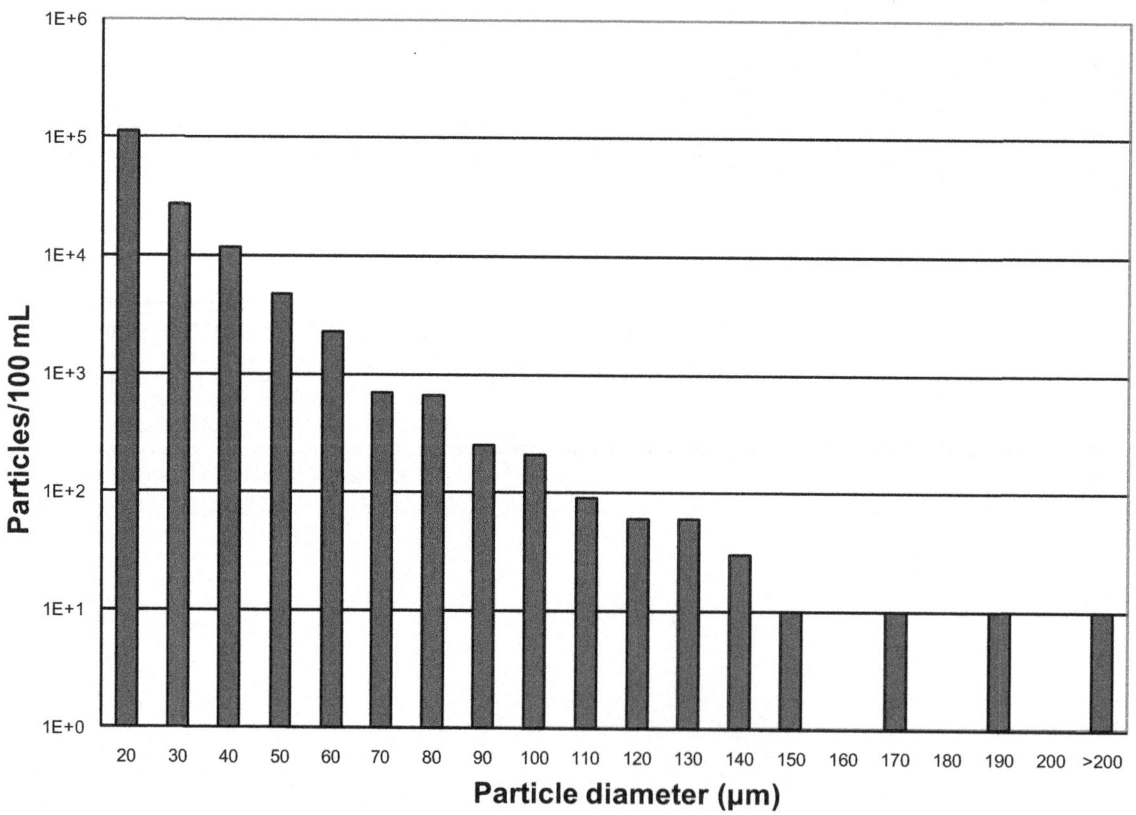


Figure B-3. Particle Size Distribution, Facility C Secondary Effluent Sample.

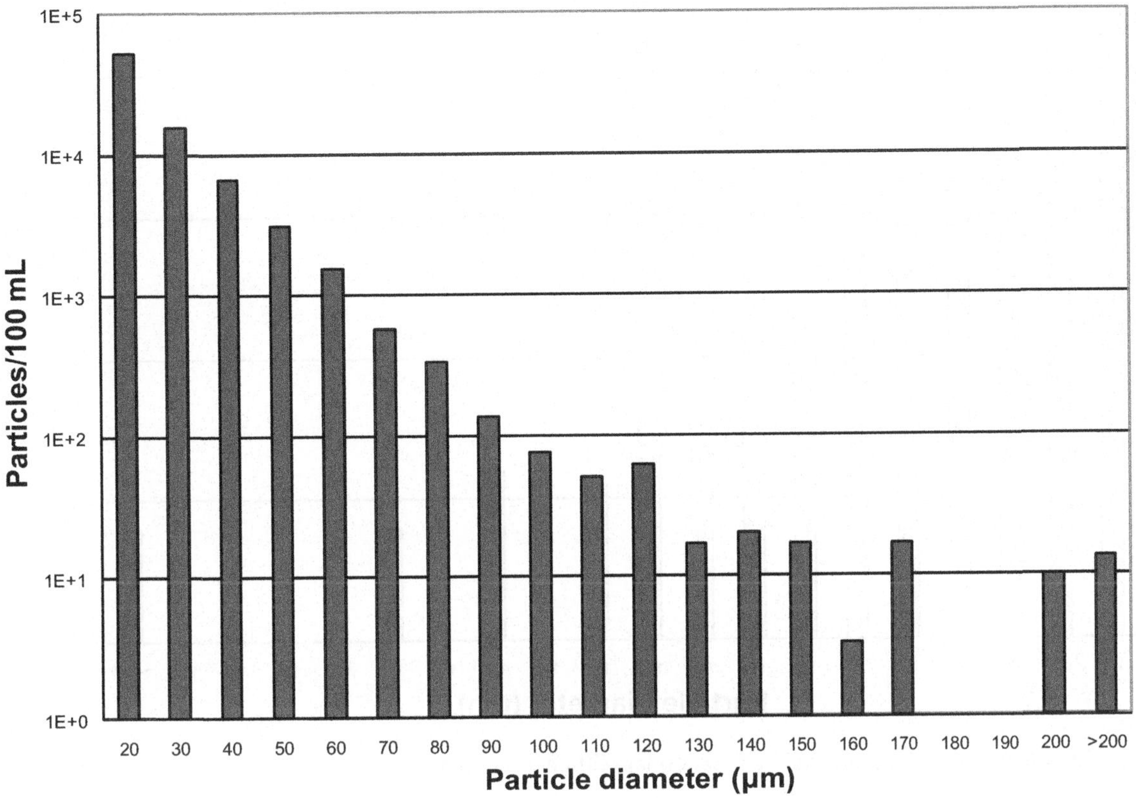


**Figure B-4. Particle Size Distribution, Facility D Secondary Effluent Sample.**

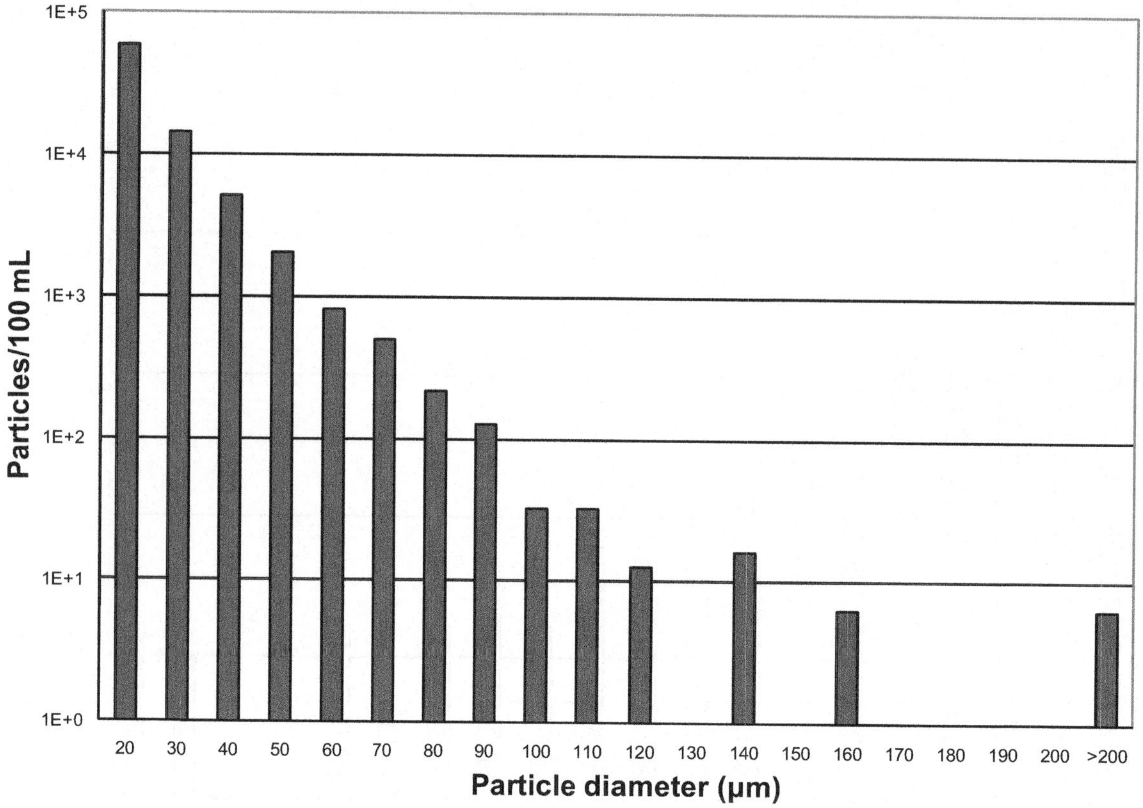


**Figure B-5. Particle Size Distribution, Facility E Secondary Effluent Sample.**

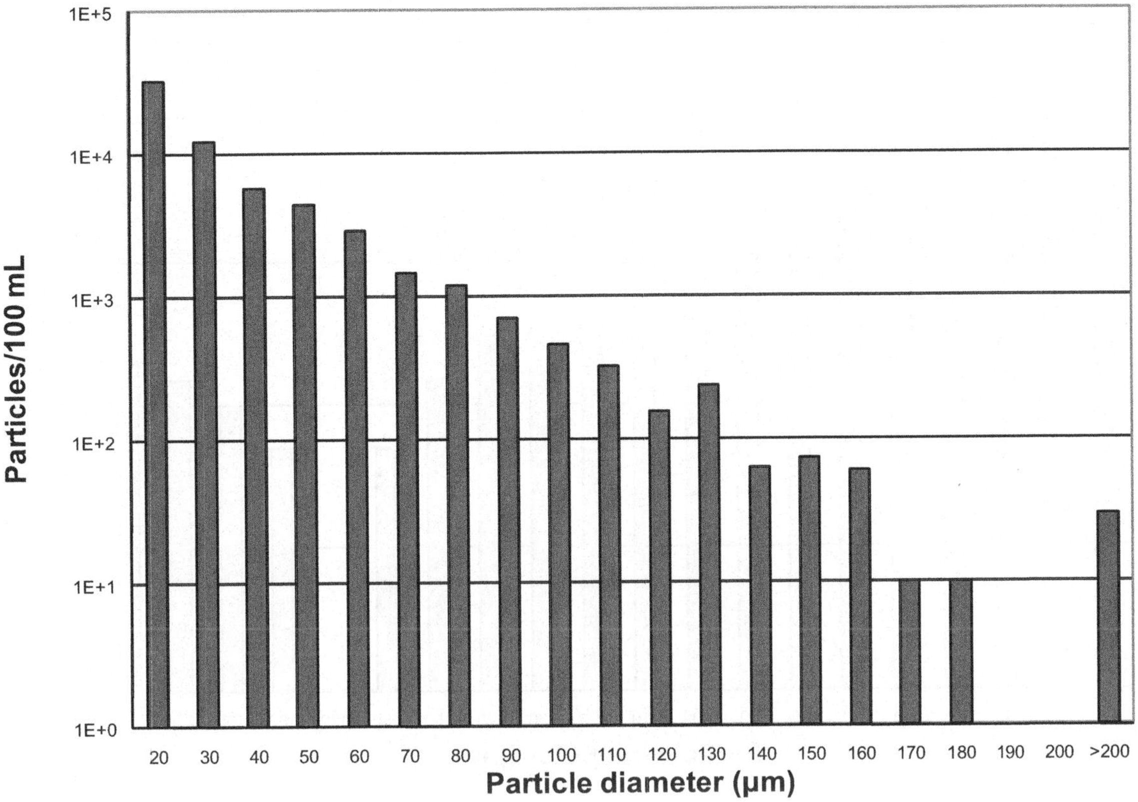


**Figure B-6. Particle Size Distribution, Facility F Secondary Effluent Sample.**

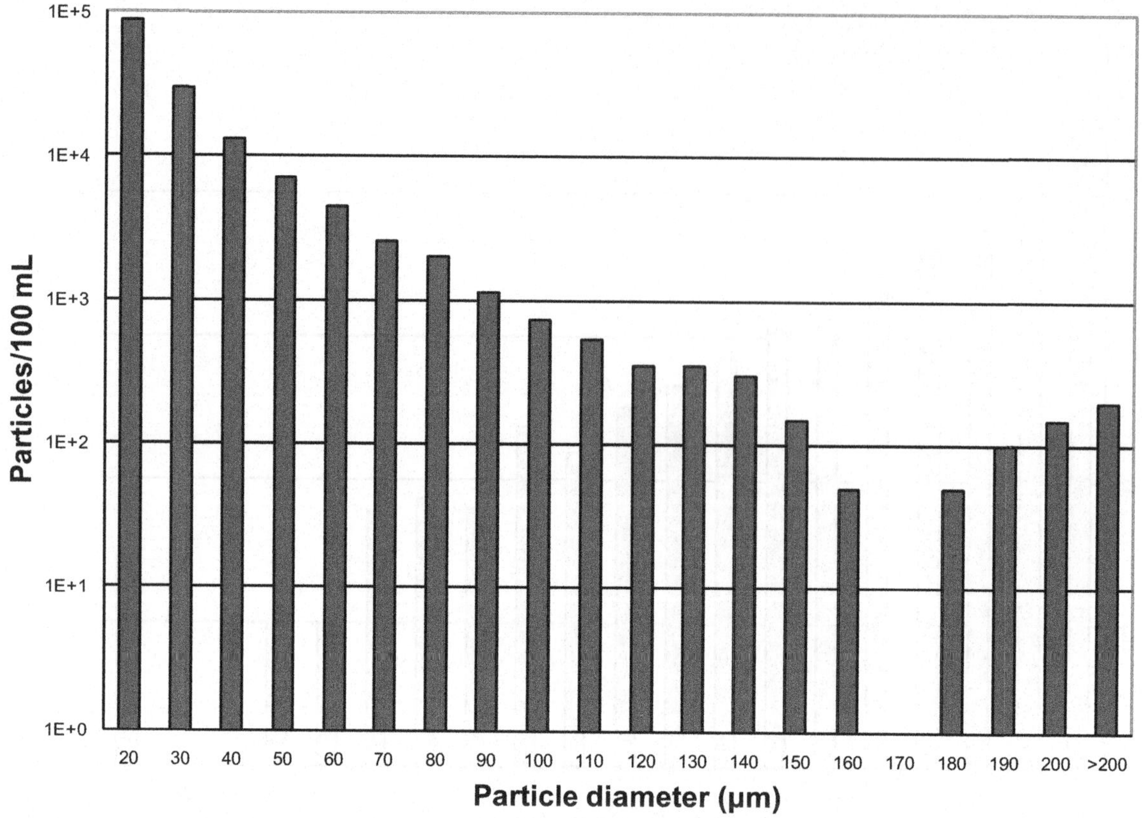


**Figure B-7. Particle Size Distribution, Facility G Secondary Effluent Sample.**

APPENDIX C

# IMAGES OF EFFLUENT WASTEWATER PARTICLES

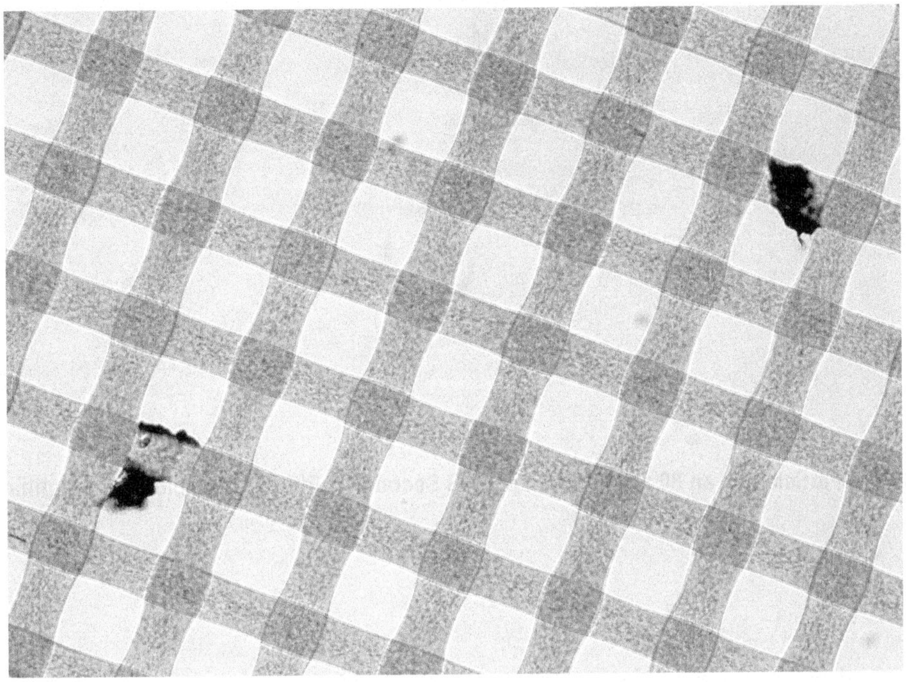

Figure C-1. Particles on an 80 μm Filter, Facility A Secondary Effluent Sample (10x Magnification).

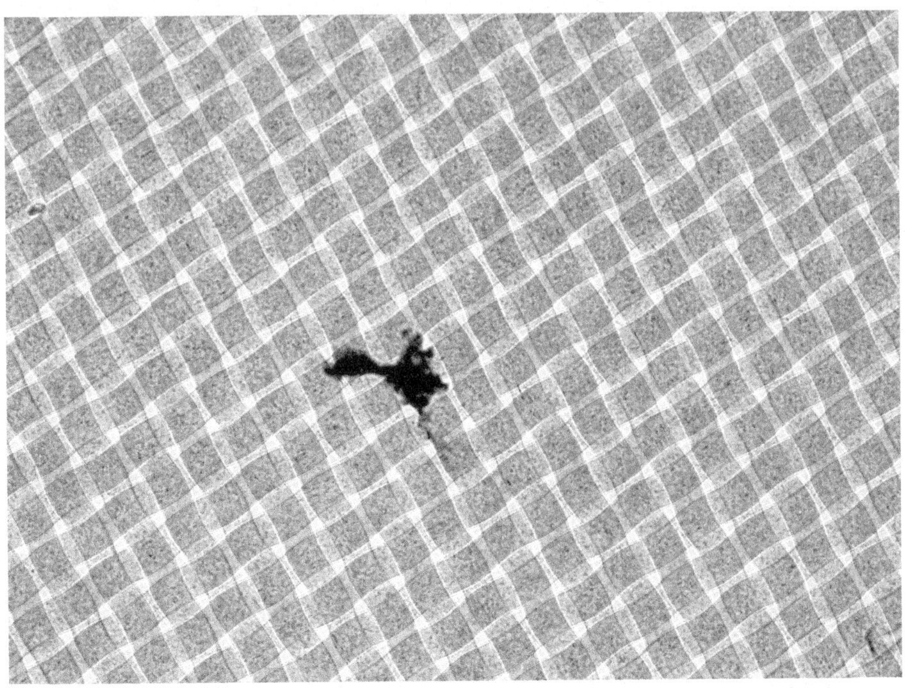

Figure C-2. Particles on an 11 μm Filter, Facility A Secondary Effluent Sample (10x Magnification).

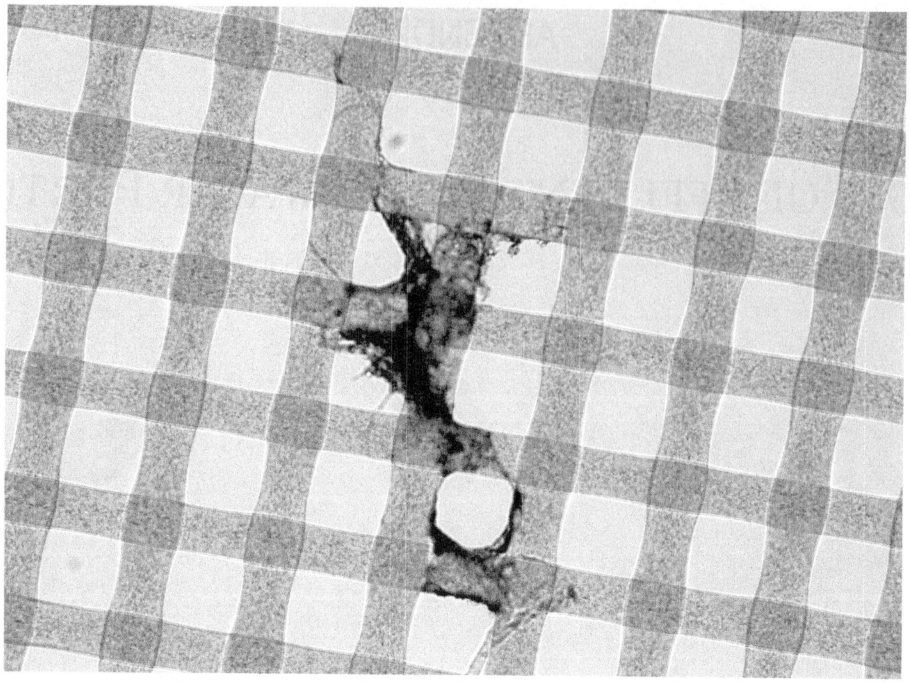

Figure C-3. Particles on an 80 μm Filter, Facility B Secondary Effluent Sample (10x Magnification).

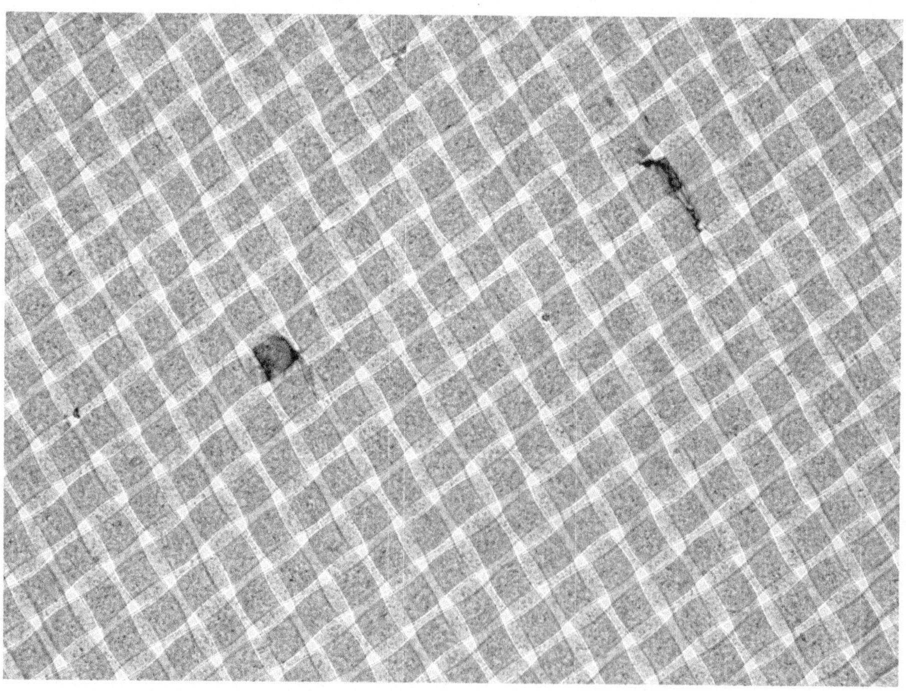

Figure C-4. Particles on an 11 μm Filter, Facility B Secondary Effluent Sample (10x Magnification).

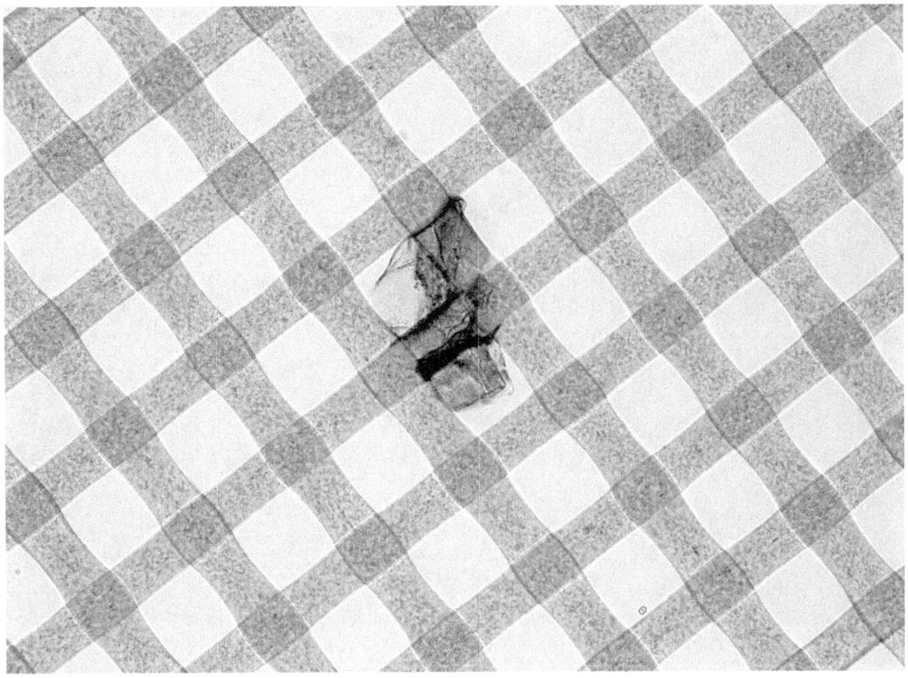

Figure C-5. Particles on an 80 μm Filter, Facility B Secondary Effluent Sample (10x Magnification).

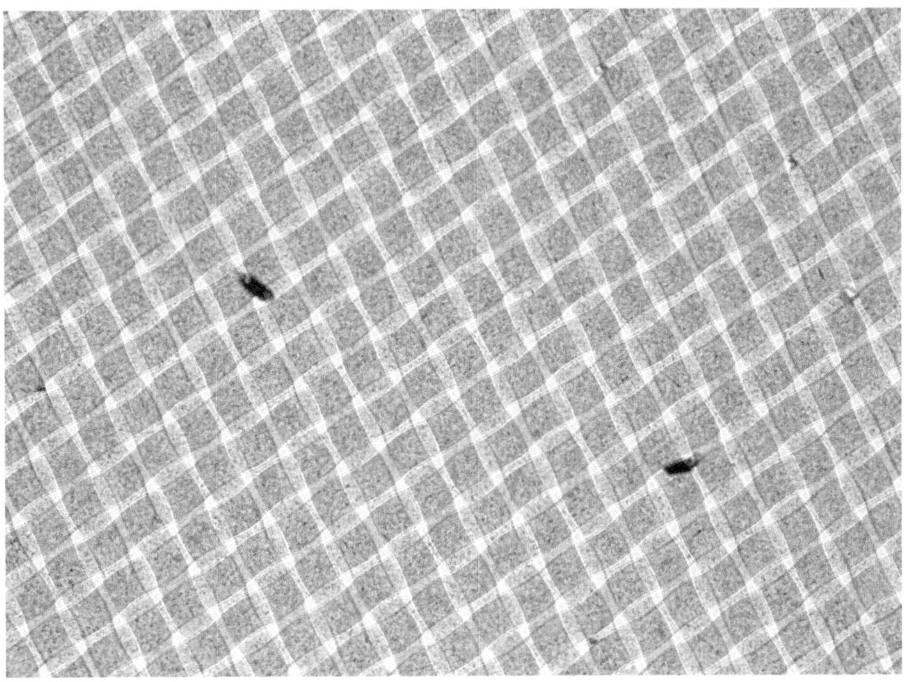

Figure C-6. Particles on an 11 μm Filter, Facility B Secondary Effluent Sample (10x Magnification).

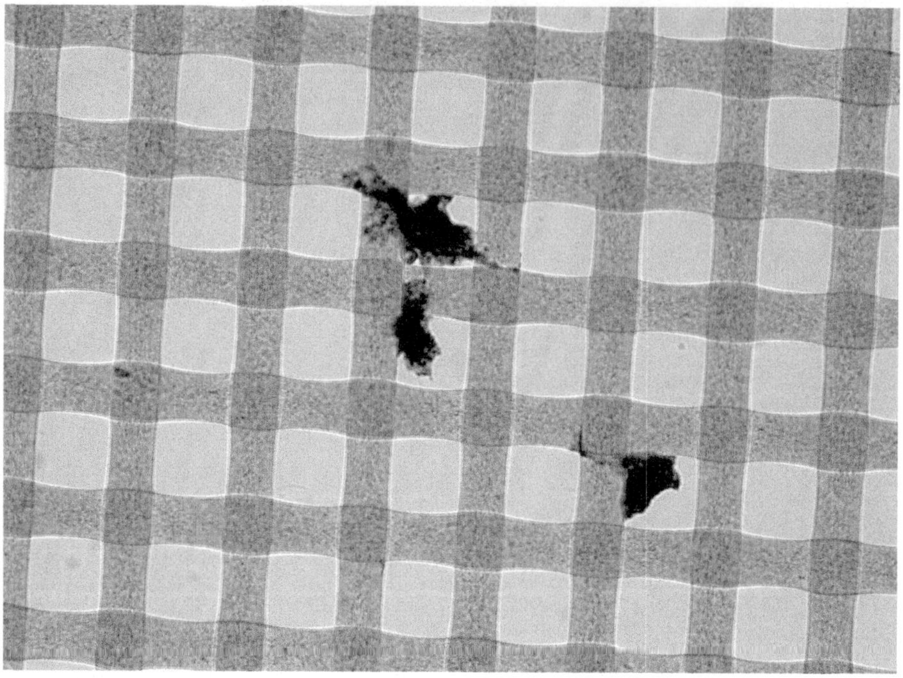

Figure C-7. Particles on an 80 μm Filter, Facility C Secondary Effluent Sample (10x Magnification).

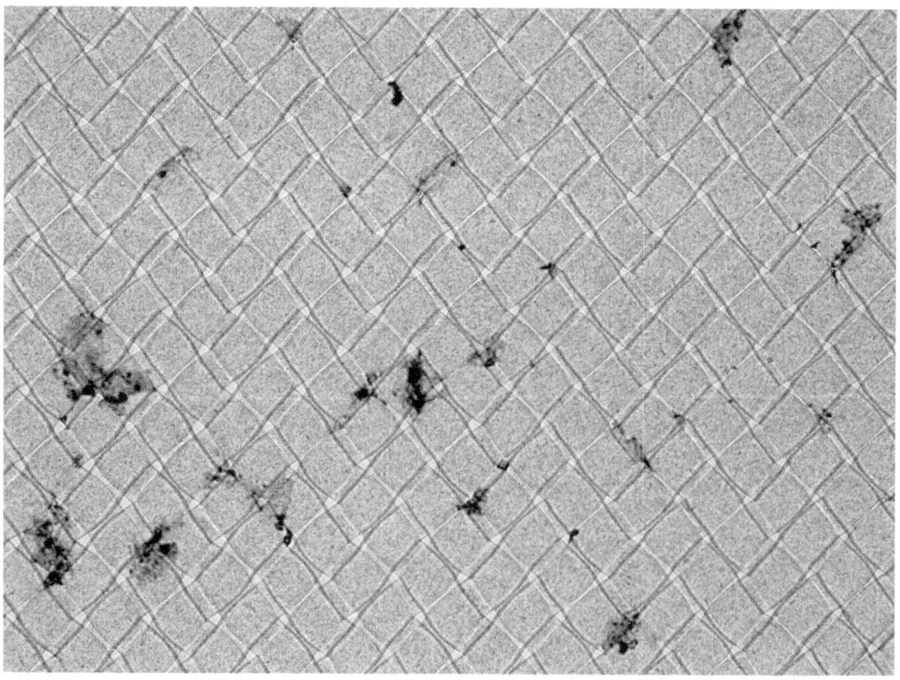

Figure C-8. Particles on an 11 μm Filter, Facility C Secondary Effluent Sample (10x Magnification).

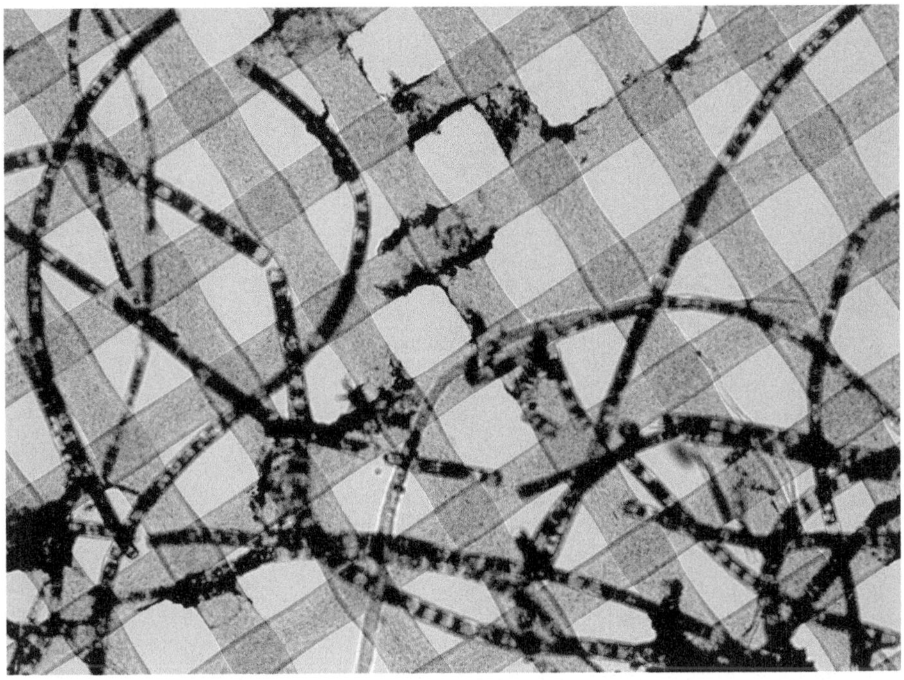

Figure C-9. Particles on an 80 μm Filter, Facility D Secondary Effluent Sample (10x Magnification).

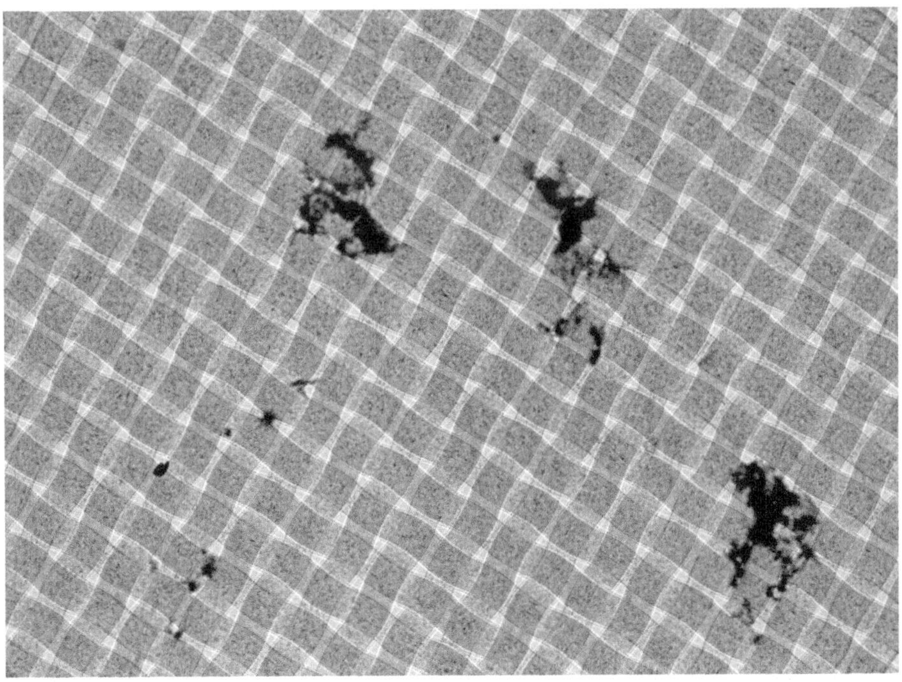

Figure C-10. Particles on an 11 μm Filter, Facility D Secondary Effluent Sample (10x Magnification).

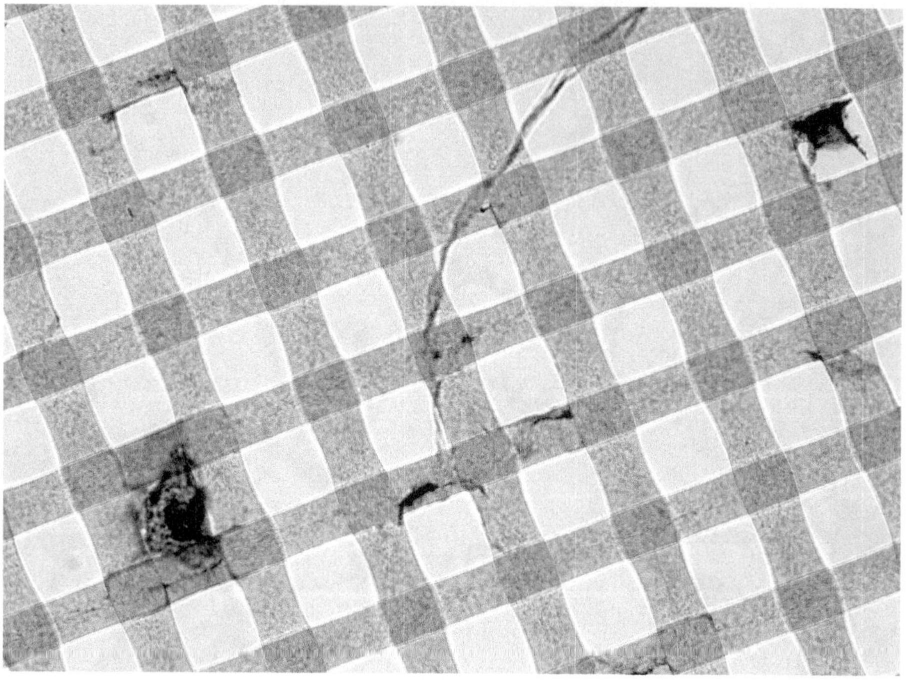

Figure C-11. Particles on an 80 μm Filter, Facility E Secondary Effluent Sample (10x Magnification).

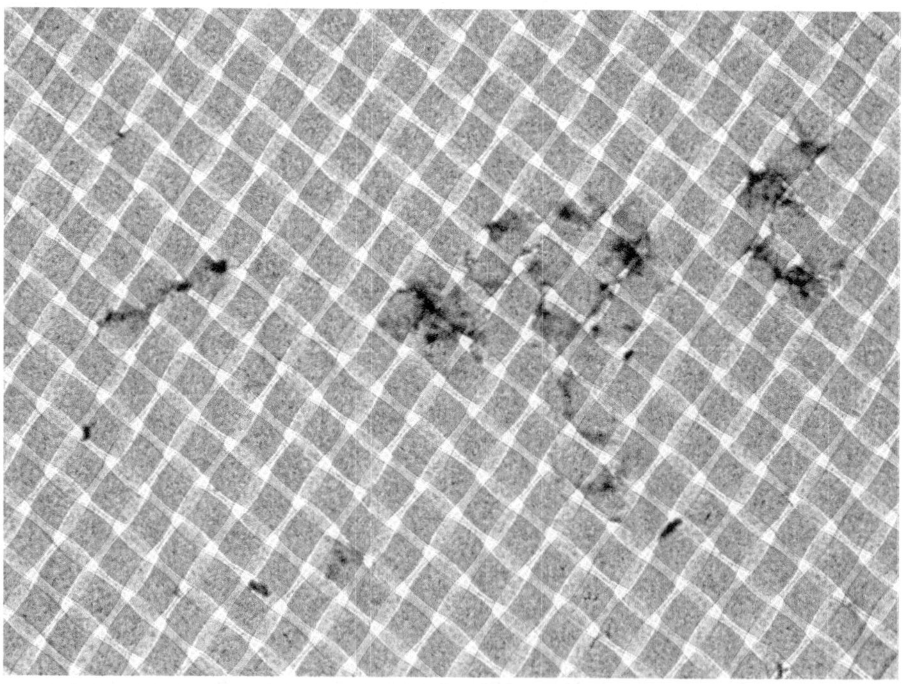

Figure C-12. Particles on an 11 μm Filter, Facility E Secondary Effluent Sample (10x Magnification).

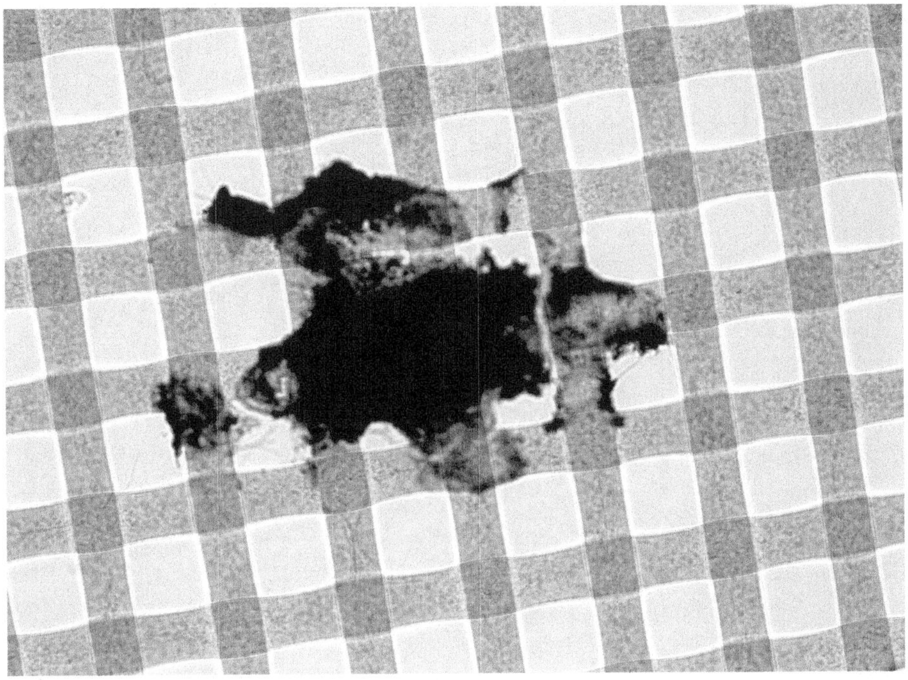

Figure C-13. Particles on an 80 μm Filter, Facility F Secondary Effluent Sample (10x Magnification).

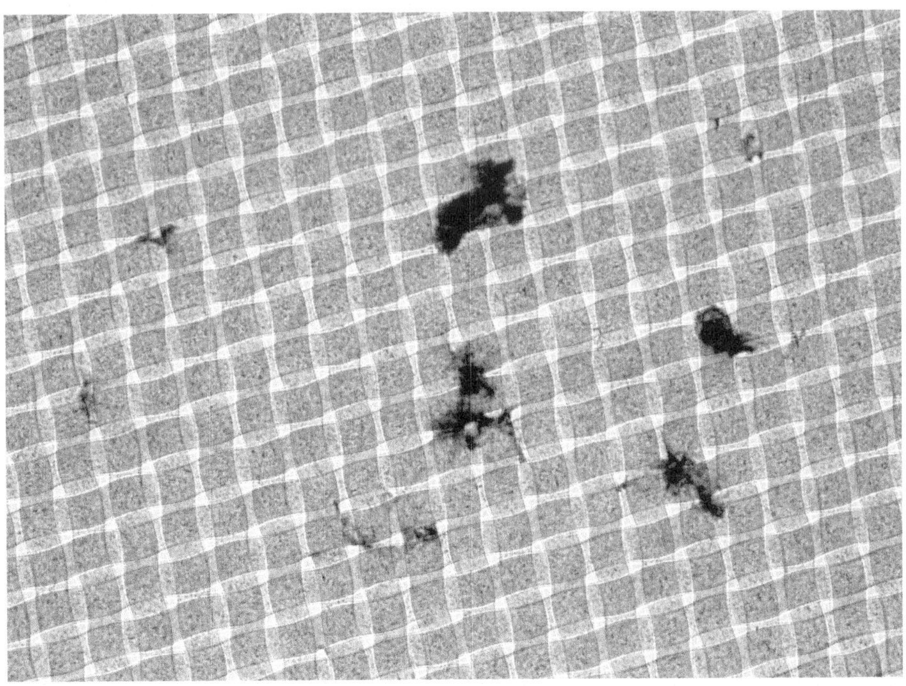

Figure C-14. Particles on an 11 μm Filter, Facility F Secondary Effluent Sample (10x Magnification).

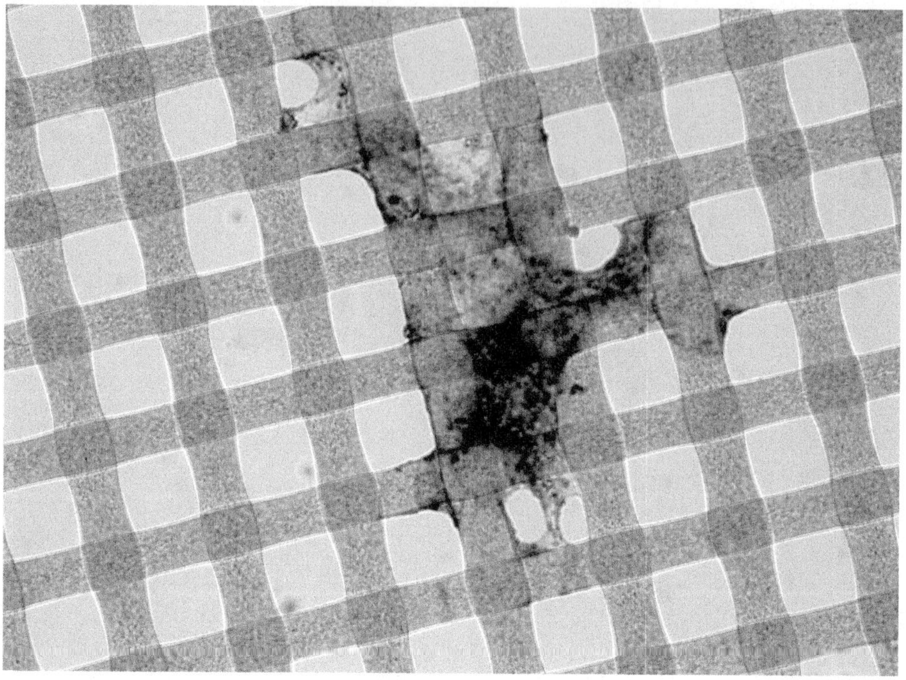

Figure C-15. Particles on an 80 μm Filter, Facility G Secondary Effluent Sample (10x Magnification).

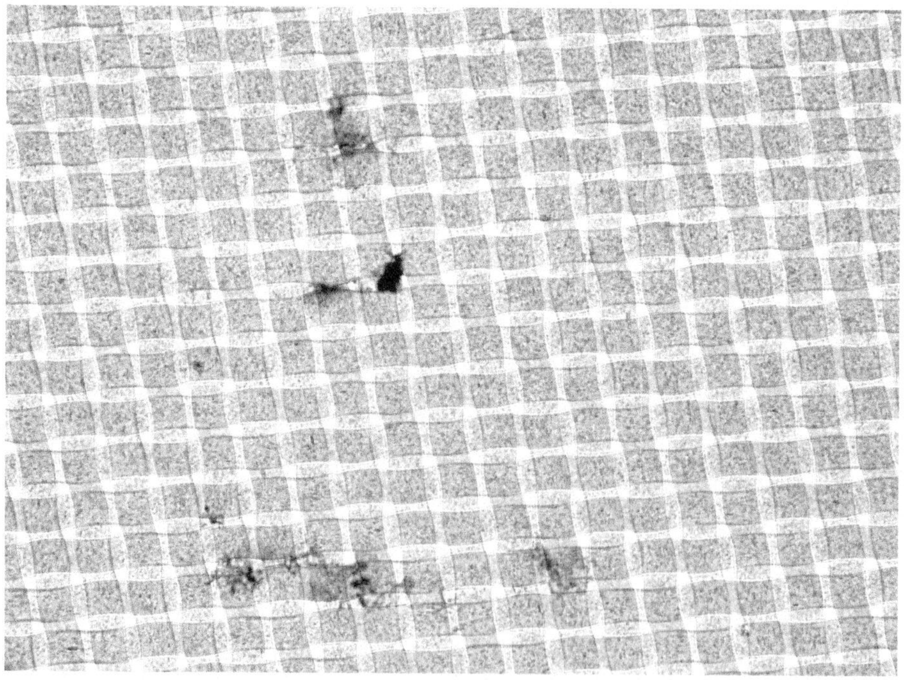

Figure C-16. Particles on an 11 μm Filter, Facility G Secondary Effluent Sample (10x Magnification).

APPENDIX D

# STEP-BY-STEP EXAMPLE: OPTIMIZATION PROTOCOLS APPLIED TO FACILITY A

The following is a summary of the application of the optimization protocols for Facility A for the first sample that was collected from Facility A (Experiment A-1). First, a secondary effluent sample that had not under gone disinfection was collected from the facility and a disinfection test was performed using the unfiltered sample (Figure D-1). The initial chlorine concentration added to the reactor in the experiment was 8 mg/L. During the disinfection test total coliform and free chlorine residual data were collected at recorded time intervals. Total coliform data were collected during the first 90-minutes (Figure D-1), and the free chlorine residual data was collected during that time and for an additional 6.5 hours (Figure D-2). The straight line in Figure D-1 is a best-fit line, as determined using a sum of least squared residual evaluation, drawn through the total coliform data in the tailing region of the inactivation curve. The best-fit line was drawn to determine the initial concentration of occluded total coliform bacteria, which is the total coliform value at the y-axis intercept.

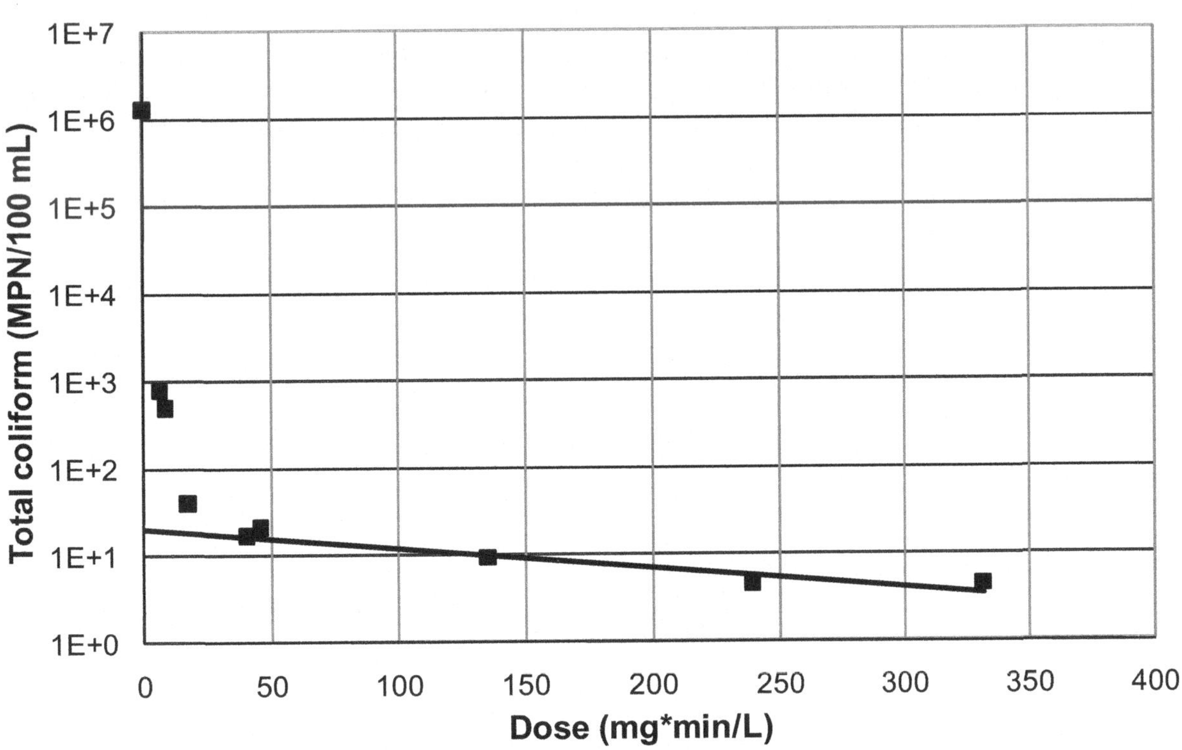


**Figure D-1. Disinfection Test Results for a Secondary Effluent Sample from Facility A.**

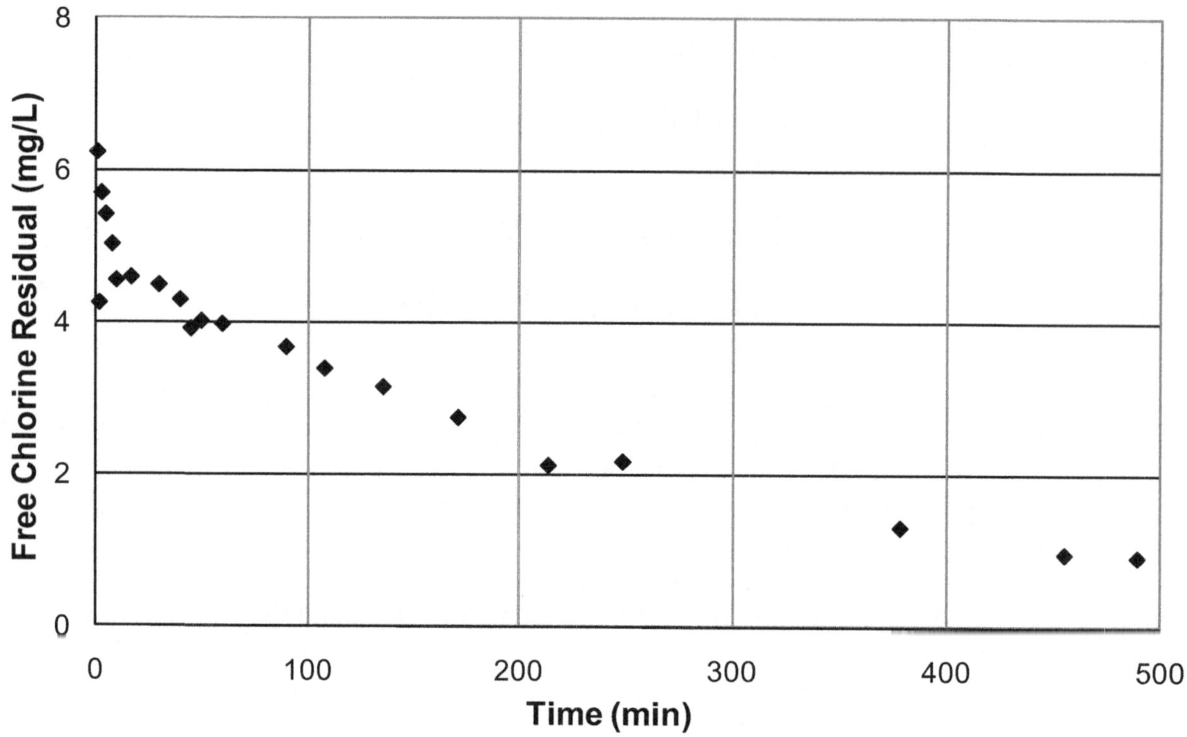


**Figure D-2. Free Chlorine Residual Data Collected During the Disinfection Test for the Secondary Effluent Sample from Facility A.**

In addition to the disinfection test performed on a secondary effluent sample, an additional disinfection test was performed on a subset of the sample that was first filtered using a 2 µm filter (Figure D-3). Data from the second disinfection test were used to calculate the first-order inactivation rate coefficient, which is equivalent to the slope of the line when the data are plotted as shown in Figure D-3. Equation 1-2 was used to calculate the inactivation rate coefficient for dispersed total coliform bacteria, which was 13.9 L/mg*min for this data set.

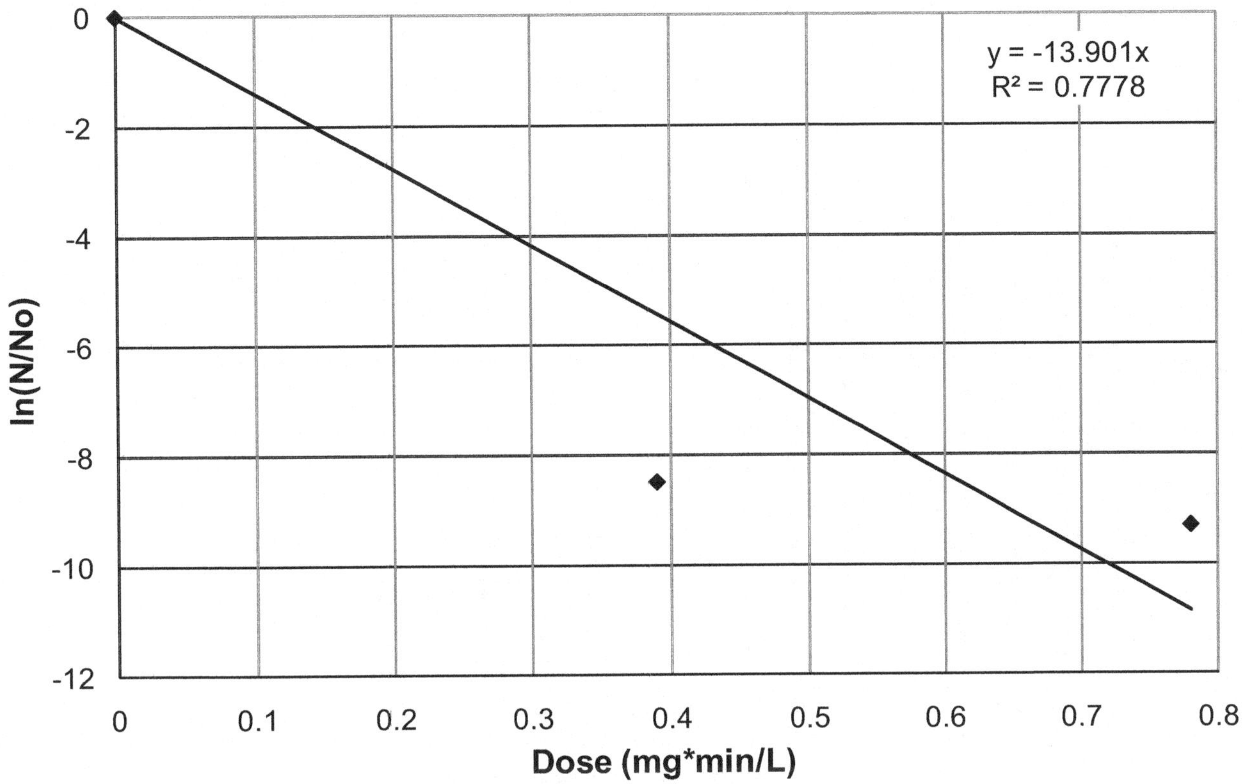


**Figure D-3. Disinfection Test Results for a Filtered (2 μm) Secondary Effluent Sample from Facility A.**

The PSD was evaluated for the secondary effluent sample (Table D-1 and Figure D-4). There were a total of 4,881 particles larger than 11 μm per 100 mL. The PSD data set contained few large particles, so the PSD was divided into four particle size classes such that in the second phase of calibration inactivation would be constrained to the largest particle size class (Table D-1). All 20 of the initially occluded total coliform bacteria were assumed to inhabit the largest particles that had a diameter of 132 μm. The particle wet density was measured for the secondary effluent particles and was 1.057 g/mL. Assuming that the density of solids was 1.4 mg/L, the calculated particle porosity was 0.86. The concentration of total coliform bacteria surviving a low dose of UV light (29) was 20 MPN/100 mL. The resulting frequency of association (F) was 0.41% for this sample.

**Table D-1 Particle Size Distribution for a Secondary Effluent Sample from Facility A.**

| Range of particle diameters (μm) | Median particle diameter | # particles/100 mL |
|---|---|---|
| 11-40 | 26 | 2941 |
| 40-80 | 50 | 1806 |
| 80-120 | 84 | 114 |
| >120 | 132 | 20 |
| **Total** | | **4881** |

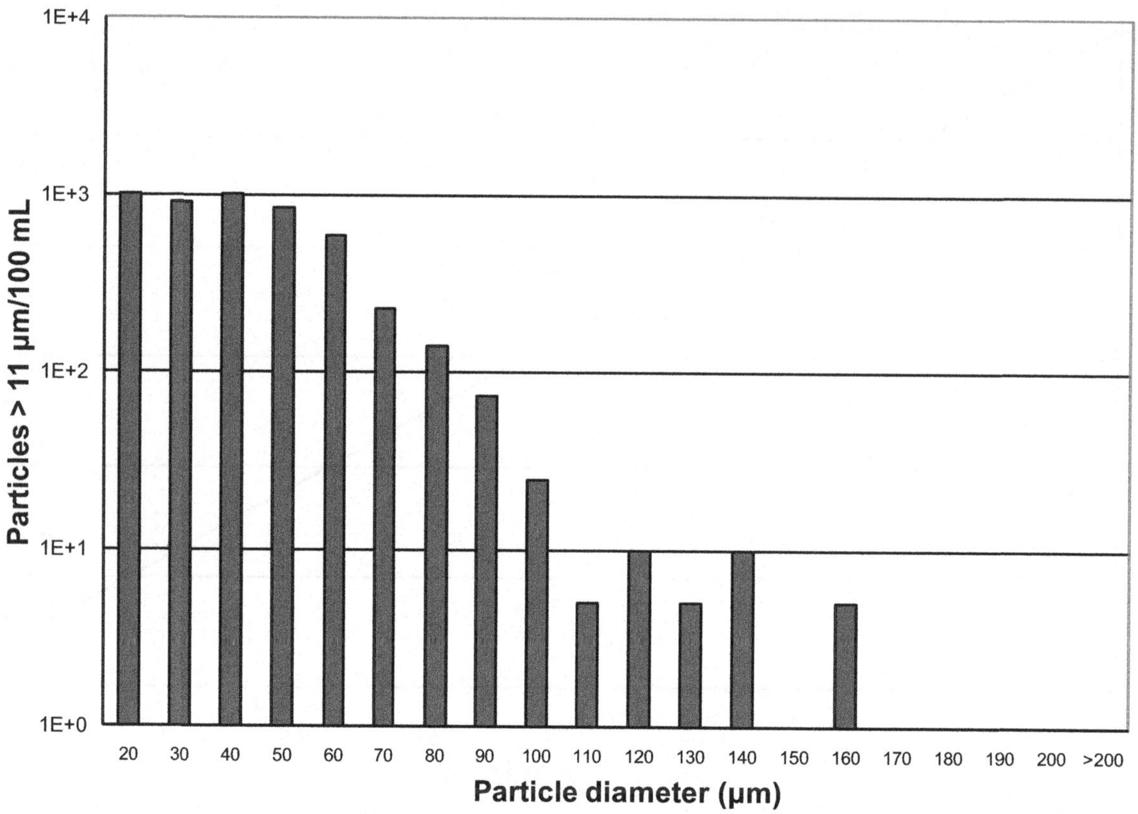


**Figure D-4. Particle Size Distribution for a Secondary Effluent Sample from Facility A.**

The observed total coliform bacteria in the tailing region of the disinfection test (Figure D-1) and the observed bulk fluid chorine (Figure D-2) were used to calibrate the model. The first three bulk fluid chlorine observations were eliminated from the calibration because it appears that this portion of the curve was part of the instantaneous chlorine demand. From viewing Figure D-2 it was determined that the instantaneous chlorine demand was approximately 3.1 mg/L, and this value yielded a good model fit for the bulk fluid chlorine data. The first four total coliform observations were not included in the calibration because these observations appeared not to be part of the tailing region of the inactivation curve. The initial concentration of occluded total coliform bacteria was determined by drawing a best-fit line through the tailing data, and was approximately 20 MPN/100 mL.

To use the data collected in laboratory studies, several calculations were necessary. To convert the radii into a dimensionless form all of the values were divided by the largest particle radius: 0.20, 0.38, 0.64, and 1.0. The number of times steps (*nts*) required was 674,000 and was determined using the following equation: $nts*\Delta\tau = \tau = tD_a/R^2$ with $\Delta\tau$=0.01, $t$=489 min, $D_a$=0.0006 $cm^2$/min, and $R$=132 µm. Occluded total coliform were assumed to inhabit the centers of the largest particles, and the corresponding intra-particle chlorine concentrations were used in the model to calculate total coliform concentrations at each time step. Several assumptions and modeling decisions were made during the calibration procedure:

- $k=k'$
- Bi=0.1
- $k_r/k_m$=100 $cm^{-1}$
- $\Delta\xi$=1/101


WERF

The model, as described by Basagaoglu et al. (2002), was modified to read in the dimensional data from the laboratory studies and calculate the dimensionless model parameters (such as the Thiele modulus). The model calculates the bulk fluid and intra-particle chlorine concentrations for the sequential time steps, and was modified to calculate the concentrations occluded total coliform bacteria and model pathogens as well. Model output consists of bulk fluid chlorine, intra-particle chlorine, and occluded microbial concentrations over time. The model output also includes a variable that consists of dimensionless time converted to dimensional time so that the model output can be compared with data collected in the laboratory studies. The model contains a variable, referred to as sampling interval, which controls the amount of output that is written to the output files. Typically, data is written to files every 10-100 time steps to avoid generating output files that are excessive large and difficult to use in graphing software programs.

The model was calibrated using parameter estimating software (UCODE) to determine the most appropriate values for $k=k'$ and $K_d$ until a best-fit of the model output for the bulk fluid chlorine and the occluded total coliform were obtained (Figure D-5). UCODE was developed by the U.S. Geological Survey, in collaboration with others, and is available electronically via the internet at no cost. The initial chlorine concentration in the bulk fluid was 8 mg/L in the disinfection test, but in the model the initial chlorine concentration in the bulk fluid ($C_{bulk,in}$) was reduced to 4.9 mg/L to provide a better fit of the model output with the data. The difference of 3.1 mg/L in initial chlorine concentrations is assumed to represent the instantaneous chlorine demand that is observed after the chlorine is introduced into the reactor. The data shown in Figure D-5 were calculated by the model and no additional data manipulation was required.

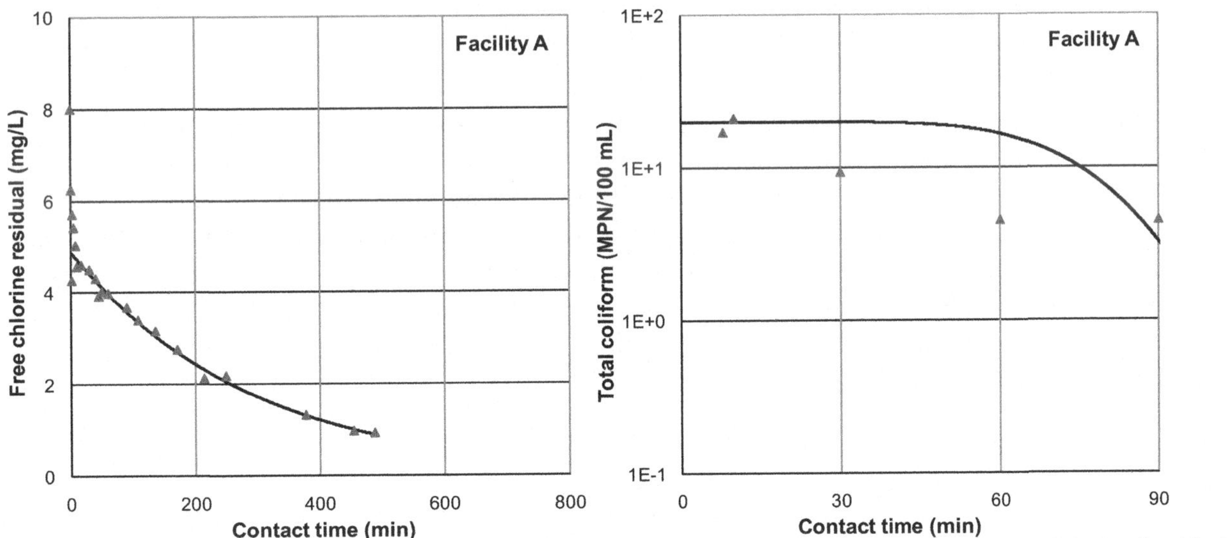


**Figure D-5. Comparison of Model Output for Chlorine Concentration in the Bulk Fluid and Observed Chlorine Residual Data Collected During the Disinfection Test of a Secondary Effluent Sample from Facility A.**

The model calibration resulted in the following parameters being selected to provide an optimal fit of the model output with the observed chlorine and total coliform data:

- $K_d$=18,670
- $k=k'$=0.003466 min$^{-1}$

It was possible to calculate the following parameters (the series of numbers refer to values for the individual particle size class, listed from the largest particle size to the smallest) for the calibrated model:

- Desorption to mass transfer at the particle surface ratio ($\eta$): 0.57, 0.36, 0.22, 0.11
- Intra-particle Thiele modulus ($\varphi$): 0.005, 0.003, 0.002, 0.001
- Thiele modulus for the external bulk fluid ($\varphi'$): 0.005

Following model calibration simulations were run using the calibrated model with different values for the initial chlorine concentration and four scenarios for particle removal (0%, 30%, 60%, and 90% removal). The result of the simulations was a matrix of surviving total coliform concentrations for a variety of initial chlorine concentrations and PSD removal rates (Table D-2).

**Table D-2. Occluded Total Coliform Concentrations Calculated Using Model Simulations.[1,2]**

| | Initial chlorine conc. (mg/L)[3] | Alterations to the PSD | | | |
|---|---|---|---|---|---|
| | | No change | 30% removal | 60% removal | 90% removal |
| Facility A | 4 | 14.3 | 10.0 | 5.7 | 1.4 |
| | 5 | 9.9 | 6.9 | 3.9 | 1.0 |
| | 6 | 6.8 | 4.8 | 2.7 | 0.7 |
| | 8 | 3.2 | 2.3 | 1.3 | 0.3 |
| | 10 | 1.5 | 1.1 | 0.6 | 0.2 |

[1]Shaded numbers represent concentrations where the particle guideline of 2/100 mL was met.
[2]Estimates were calculated assuming total coliform occluded in larger particles prior to smaller particles.
[3]Initial chlorine concentration is the initial model concentration plus the instantaneous chlorine demand.

For highly treated wastewaters, a regulatory limit for total coliform bacteria of 2 MPN/100 mL is often imposed, which is the detection limit for the total coliform test. As shown in Table D-3, this regulatory limit was met in 9 of the 20 treatment scenarios (shaded areas). Where the removal rate of particles was high, 90%, the treatment objective was met. Where the initial chlorine concentration was high, 10 mg/L, the treatment objective was also met. It was also possible to meet the treatment objective using a combination of particle removal (60%) and an initial chlorine concentration of 8 mg/L. The data in Table D-2 are also shown in a contour plot in Figure D-6.

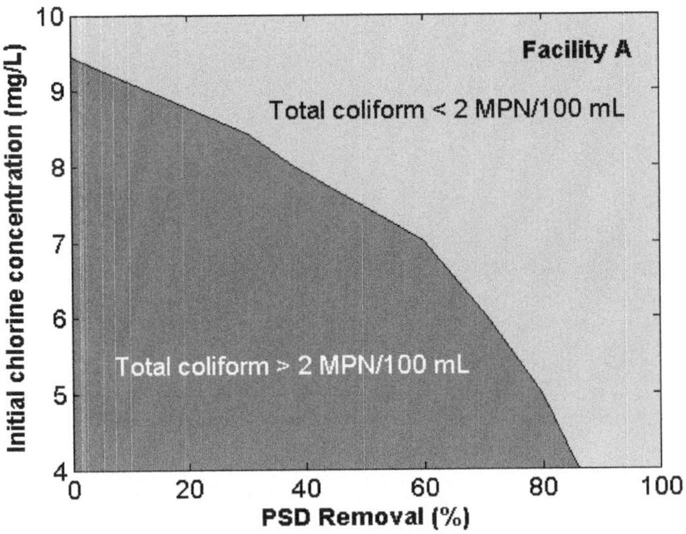


**Figure D-6. Combinations of Particle Removal and Initial Total Chlorine Concentrations That Result in Occluded Total Coliform Concentrations Greater Than and Less Than 2 MPN/100 mL, as Predicted Using Model Output for Facility A.**

The particle guidelines that were calculated for Facility A are shown in Table D-3 (see Chapters 2.0 and 6.0 for more detail). When the PSD data for Facility A (Table D-1) are compared with the particle guidelines, it is clear that the total particle count of 4881/100 mL is less than the second two particle guidelines in Table D-3, but higher than the first particle guideline that is based on drinking water and reclaimed water standards. To reliably meet the treatment goal that the first particle guideline is based on (2 MPN/100 mL), the disinfection system must be designed and operated to inactivate occluded total coliform bacteria.

**Table D-3. Particle Guidelines (# / 100 mL) for Facility A Based on Indicator Organisms.[1]**

| Microorganism basis | Exposure route | Facility A |
|---|---|---|
| *E. coli* and coliforms | Drinking water, reclaimed water | 488 |
| Enterococci | Recreational fresh water | 8049 |
| Enterococci | Recreational marine water | 8537 |

[1]Particle guidelines were calculated using Equation 2-1. Measured values of F were used and P was assumed to be one. Particle guidelines represent the effluent limits for particles to meet the regulatory requirements for the indicated microorganisms, assuming that the dispersed target microorganisms are completely inactivated and that the occluded target microorganisms are not inactivated in the disinfection system.

The optimization protocols completed (see results in Table D-2 and Figure D-6) present an approach to determining how the standards can be met. Managers of treatment facilities, such as Facility A, are left to decide whether they should make changes and improvements to their filtration systems or to their disinfection systems. Typically, a combination of filtration and disinfection is used to meet stringent standards such as a microbial limit of 2 total coliform per 100 mL.

# REFERENCES

American Public Health Association (APHA). (1998). *Standard Methods for the Examination of Water and Wastewater*, 20th ed. American Public Health Association, Washington, D.C.

Asano, T., Leong, L.Y.C., Rigby, M.G., and Sakaji, R.H. (1992). "Evaluation of the California wastewater reclamation criteria using enteric virus monitoring data." *Water Science and Technology*, 26(7-8), 1513-1524.

Asano, T., Metcalf & Eddy, Burton, F.L., Leverenz, H.L., Tsuchihashi, R., and Tchobanoglous, G. (2007). *Water Reuse*, McGraw-Hill, New York, NY.

Baker, K.H., Hegarty, J.P., Redmond, B., Reed, N.A., and Herson, D.S. (2002). "Effect of oxidizing disinfectants (chlorine, monochloramine, and ozone) on *Helicobacter pylori*." *Applied and Environmental Microbiology*, 68(2), 981-984.

Barbeau, B., Boulos, L., Desjardins, R., Coallier, J., and Prevost, M. (1999). "Examining the use of aerobic spore-forming bacteria to assess the efficiency of chlorination." *Water Research*, 33(13), 2941-2948.

Basagaoglu, H., McCoy, B.J., Ginn, T.R., Loge, F.J., and Dietrich, J.P. (2002). "A diffusion limited sorption kinetics model with polydispersed particles of distinct sizes and shapes." *Advances in Water Resources*, 25(7), 755-772.

Berman, D., Rice, E.W., and Hoff, J.C. (1988). "Inactivation of particle-associated coliforms by chlorine and monochloramine." *Applied and Environmental Microbiology*, 54(2), 507-512.

Bitton, G. (2005). *Wastewater Microbiology*, John Wiley and Sons, Hoboken, NJ.

Borst, M. and Selvakumar, A. (2003). "Particle-associated microorganisms in stormwater runoff." *Water Research*, 37(1), 215-223.

Braunstein, J.L., Loge, F.J., Tchobanoglous, G., and Darby, J.L. (1996). "Ultraviolet disinfection of filtered activated sludge effluent for reuse applications." *Water Environment Research*, 68(2), 152-161.

Brown and Caldwell. (2005). Miami-Dade Water and Sewer Department South District Wastewater Treatment Facility Disinfection Demonstration: Volume II Exhibits, Version 2.0, Miami, FL.

California Department of Public Health (CA DPH). (2001). "California Health Laws Related to Recycled Water: The Purple Book." Prepared by: Division of Drinking Water and Environmental Management. June 2001 edition, Carpinteria, CA.

California Department of Public Health (CA DPH). (2008). "Treatment Technology Report for Recycled Water." Prepared by: Division of Drinking Water and Environmental Management, Carpinteria, CA.

Cantwell, R.E., Hofmann, R., and Templeton, M.R. (2008). "Interactions between humic matter and bacteria when disinfecting water with UV light." *Journal of Applied Microbiology*, 105(1), 25-35.

Camper, A.K., Lechevallier, M.W., Broadaway, S.C., and McFeters, G.A. (1985). "Evaluation of procedures to desorb bacteria from granular activated carbon." *Journal of Microbiological Methods*, 3(3-4), 187-198.

Caron, E., Cheurefils, G., Barbeau, B., Payment, P., and Prevost, M. (2007). "Impact of microparticles on UV disinfection of indigenous aerobic spores." *Water Research*, 41(19), 4546-4556.

Chao, M.S. (1968). "The diffusion coefficients of hypochlorite, hypochlorous acid, and chlorine in aqueous media by chronopotentiometry." Journal Electrochemical Society: Electrochemical Science, 115(11), 1172-1174.

County Sanitation Districts of Los Angeles County (CSDLAC). (1977). "Pomona Virus Study, Final Report." Prepared for Calif. State Water Resources Control Board, Sacramento, Calif., and U.S. EPA, Washington, D.C.

Dietrich, J.P., Basagaoglu, H., Loge, F.J., and Ginn, T.R. (2003). "Preliminary assessment of transport processes influencing the penetration of chlorine into wastewater particles and the subsequent inactivation of particle-associated organisms." *Water Research*, 37(1), 139-149.

Dietrich, J.P., Loge, F.J., Ginn, T.R., and Basagaoglu, H. (2007). "Inactivation of particle-associated microorganisms in wastewater disinfection: Modeling of ozone and chlorine reactive diffusive transport in polydispersed suspensions." *Water Research*, 41(10), 2189-2201.

Emerick, R.W., Loge, F.J., Ginn, T., and Darby, J.L. (2000). "Modeling the inactivation of particle-associated coliform bacteria." *Water Environment Research*, 72(4), 432-438.

Emerick, R.W., Loge, F.J., Thompson, D., and Darby, J.L. (1999). "Factors influencing ultraviolet disinfection performance part II: Association of coliform bacteria with wastewater particles." *Water Environment Research*, 71(6), 1178-1187.

Gyurek, L.L., Finch, G.R., and Belosevic, M. (1997). "Modeling chlorine inactivation requirements of *Cryptosporidium parvum* oocysts." *Journal of Environmental Engineering-ASCE*, 123(9), 865-875.

Haas, C.N., Rose, J.B., and Gerba, C.P. (1999). *Quantitative Microbial Risk Assessment*, Hoboken, NJ.

Harriott, P. (1962). "Mass transfer to particles: Part I. Suspended in agitated tanks." American Institue of Chemical Engineers Journal, 8(1), 93-101.

Hejkal, T.W., Smith, E.M., and Gerba, C.P. (1984). "Seasonal occurrence of rotavirus in sewage." *Applied and Environmental Microbiology*, 47(3), 588-590.

Hejkal, T.W., Wellings, F.M., Larock, P.A., and Lewis, A.L. (1979). "Survival of poliovirus within organic solids during chlorination." *Applied and Environmental Microbiology*, 38(1), 114-118.

Herson, D.S., McGonigle, B., Payer, M.A., and Baker, K.H. (1987). "Attachment as a factor in the protection of *Enterobacter cloacae* from chlorination." *Applied and Environmental Microbiology*, 53(5), 1178-1180.

Hsu, C.L. (2003). "Biot Number." *Encyclopedia of Agricultural, Food, and Biological Engineering*, D.R. Heldman, ed, Marcel Dekker, Inc., New York.

Irving, L.G. and Smith, F.A. (1981). "One-year survey of enteroviruses, adenoviruses, and reoviruses isolated from effluent at an activated-sludge purification plant." *Applied and Environmental Microbiology*, 41(1), 51-59.

Jenkins, D., Richard, M.G., and Daigger, G.T. (2003). *Manual on the Causes and Control of Activated Sludge Bulking, Foaming, and Other Solids Separation Problems*, CRC Press LLC, London, UK.

John, D.E., Haas, C.N., Nwachuku, N., and Gerba, C.P. (2005). "Chlorine and ozone disinfection of *Encephalitozoon intestinalis* spores." *Water Research*, 39(11), 2369-2375.

Jolis, D., Hirano, R., and Pitt, P. (1999). "Tertiary treatment using microfiltration and UV disinfection for water reclamation." *Water Environment Research*, 71(2), 224-231.

Kaminski, I., Vescan, N., and Adin, A. (1997). "Particle size distribution and wastewater filter performance." *Water Science and Technology*, 36(4), 217-224.

Kobler, D. and Boller, M. (1997). "Particle removal in different filtration systems for tertiary wastewater treatment - A comparison." *Water Science and Technology*, 36(4), 259-267.

Lambert, R.J.W. and Johnston, M.D. (2000). "Disinfection kinetics: a new hypothesis and model for the tailing of log-survivor/time curves." *Journal of Applied Microbiology*, (88), 907-913.

LeChevallier, M.W., Cawthon, C.D., and Lee, R.G. (1988). "Inactivation of biofilm bacteria." *Applied and Environmental Microbiology*, 54(10), 2492-2499.

Li, D.H. and Ganczarczyk, J.J. (1987). "Stroboscopic determination of settling velocity, size and porosity of activated sludge particles." *Water Research*, 21(3), 257-262.

Loge, F.J., Bourgeous, K., Emerick, R.W., and Darby, J.L. (2001). "Variations in wastewater quality parameters influencing UV disinfection performance: Relative impact of filtration." *Journal of Environmental Engineering-ASCE*, 127(9), 832-837.

Loge, F.J., Emerick, R.W., Ginn, T.R., and Darby, J.L. (2002). "Association of coliform, bacteria with wastewater particles: impact of operational parameters of the activated sludge process." *Water Research*, 36(1), 41-48.

Loge, F.J., Emerick, R.W., Thompson, D.E., Nelson, D.C., and Darby, J.L. (1999). "Factors influencing ultraviolet disinfection performance part I: Light penetration to wastewater particles." *Water Environment Research*, 71(3), 377-381.

Madge, B.A. and Jensen, J.N. (2006). "Ultraviolet disinfection of fecal coliform in municipal wastewater: Effects of particle size." *Water Environment Research*, 78(3), 294-304.

Mamane-Gravetz, H. and Linden, K.G. (2004). "UV disinfection of indigenous aerobic spores: implications for UV reactor validation in unfiltered waters." *Water Research*, 38(12), 2898-2906.

McCullough, N.B. and Eisele, C.W. (1951). "Experimental human salmonellosis .3. Pathogenicity of strains of Salmonella newport, Salmonella derby, and Salmonella bareilly obtained from spray-dried whole egg." *Journal of Infectious Diseases*, 89(3), 209-213.

National Research Council (NRC). (1998). *Issues in potable reuse*, National Academy Press, Washington, D.C.

Olivieri, V. P. (1985). "Human pathogens, disinfection, and chlorine." Water chlorination, chemistry, environmental impact and health effects, R.L. Jolley, R.J. Bull, W.P. Davis, S. Katz, M.H. Roberts, Jr., and V.A. Jacobs, eds., Lewis Publishers, Inc., Chelsea, MI, 5-18.

Ormeci, B. and Linden, K.G. (2002). "Comparison of UV and chlorine inactivation of particle and non-particle associated coliform." *Water Science and Technology: Water Supply*, 2(5-6), 403-410.

Ormeci, B. and Linden, K.G. (2005). "Comparison of physical and chemical methods for extraction of coliform from wastewater particles and particles." *Environmental Engineering Science*, 22(4), 459-471.

Payment, P., Tremblay, M., and Trudel, M. (1985). "Relative resistance to chlorine of poliovirus and coxsackievirus isolates from environmental sources and drinking-water." *Applied and Environmental Microbiology*, 49(4), 981-983.

Parker, J.A. and Darby, J.L. (1995). "Particle-associated coliform in secondary effluents - shielding from ultraviolet light disinfection." *Water Environment Research*, 67(7), 1065-1075.

Payment, P., Berte, A., Prevost, M., Menard, B., and Barbeau, B. (2000). "Occurrence of pathogenic microorganisms in the Saint Lawrence River (Canada) and comparison of health risks for populations using it as their source of drinking water." *Canadian Journal of Microbiology*, 46(6), 565-576.

Poeter, E.P., Hill, M.C., Banta, E.R., Mehl, S., and Christensen, S. (2005). "UCODE_2005 and six other computer codes for universal sensitivity analysis, calibration, and uncertainty evaluation." U.S.G.S. Techniques and Methods 6-A11, Reston, VA.

Rice, E.W., Clark, R.M., and Johnson, C.H. (1999). "Chlorine inactivation of *Escherichia coli* O157 : H7." *Emerging Infectious Diseases*, 5(3), 461-463.

Rice, E.W., Adcock, N.J., Sivaganesan, M., Brown, J.D., Stallknecht, D.E., and Swayne, D.E. (2007). "Chlorine inactivation of highly pathogenic avian influenza virus (H5N1)." *Emerging Infectious Diseases*, 13(10), 1568-1570.

Ridgway, H.F. and Olson, B.H. (1982). "Chlorine resistance patterns of bacteria from 2 drinking water distribution systems." *Applied and Environmental Microbiology*, 44(4), 972-987.

Rose, J.B., Dickson, L.J., Farrah, S.R., and Carnahan, R.P. (1996). "Removal of pathogenic and indicator microorganisms by a full-scale water reclamation facility." *Water Research*, 30(11), 2785-2797.

Rose, J.B., Farrah, S.R., Harwood, V.J., Levine, A.D., Lukasik, J., Menendez, P., and Scott, T.M. (2004). "Reduction of Pathogens, Indicator Bacteria, and Alternative Indicators by Wastewater Treatment and Reclamation Processes." WERF, Project No. 00-PUM-2T, Alexandria, VA.

Rose, J.B., Haas, C.N., and Regli, S. (1991). "Risk assessment and control of waterborne giardiasis." *American Journal of Public Health*, 81(6), 709-713.

Rose, J.B., Huffman, D.E., Riley, K., Farrah, S.R., Lukasik, J.O., and Hamann, C.L. (2001). "Reduction of enteric microorganisms at the Upper Occoquan Sewage Authority Water Reclamation Plant." *Water Environment Research*, 73(6), 711-720.

Rose, L.J., Rice, E.W., Hodges, L., Peterson, A., and Arduino, M.J. (2007). "Monochloramine inactivation of bacterial select agents." *Applied and Environmental Microbiology*, 73(10), 3437-3439.

Roseberry, A.M. and Burmaster, D.E. (1992). "Lognormal distributions for water intake by children and adults." *Risk Analysis*, 12(1), 99-104.

Shin, G.A., and Sobsey, M.D. (2008). "Inactivation of norovirus by chlorine disinfection of water." *Water Research*, 42(17), 4562-4568.

Snoeyink, V.L. and Jenkins, D. (1980). *Water Chemistry*, John Wiley, New York.

Sobsey, M.D. (1989). "Inactivation of health-related microorganisms in water by disinfection processes." *Water Science and Technology*, 21(3), 179-195.

Sobsey, M.D., Fuji, T., and Hall, R.M. (1991). "Inactivation of cell-associated and dispersed hepatitis A virus in water." *Journal American Water Works Association*, 83(11), 64-67.

Sobsey, M.D., Fuji, T., and Shields, P.A. (1988). " Inactivation of Hepatitis A virus and model viruses in water by free chlorine and monochloramine." *Water Science and Technology*, 20(11/12), 385-391.

Stewart, P.S., and Raquepas, J.B. (1995). "Implications of reaction-diffusion theory for the disinfection of microbial biofilms by reactive antimicrobial agents." *Chemical Engineering Science*, 50(19), 3099-3104.

Tanaka, H., Asano, T., Schroeder, E.D., and Tchobanoglous, G. (1998). "Estimating the safety of wastewater reclamation and reuse using enteric virus monitoring data." *Water Environment Research*, 70(1), 39-51.

Tchobanoglous, G., Burton, F.L., and Stensel, H.D. (2003). *Wastewater Engineering: Treatment and Reuse,* McGraw-Hill, New York.

Templeton, M.R., Andrews, R.C., and Hofmann, R. (2005). "Inactivation of particle-associated viral surrogates by ultraviolet light." *Water Research*, 39(15), 3487-3500.

Thurston-Enriquez, J.A., Haas, C.N., Jacangelo, J., and Gerba, C.P. (2005). "Inactivation of enteric adenovirus and feline calicivirus by chlorine dioxide." Applied and Environmental Microbiology, 71(6), 3100-3105.

U.S. Environmental Protection Agency. (1986). Ambient Water Quality Criteria for Bacteria, EPA440/5-84-002, Washington, D.C.

U.S. Environmental Protection Agency. (1989). National Primary Drinking Water Regulations: Filtration; Disinfection; Turbidity, *Giardia lamblia*, Viruses, *Legionella*, and Heterotrophic bacteria, Final Rule, Federal Register, 54:27486 (June 28, 1989), Washington, D.C.

U.S. Environmental Protection Agency. (2002). National Primary Drinking Water Regulations: Maximum contaminant level goals for microbiological contaminants, 40CFR141.52 (July 1, 2002), Washington, D.C.

U.S. Environmental Protection Agency. (2004). Guidelines for Water Reuse, EPA/625/R-04/108, Washington, D.C.

U.S. Environmental Protection Agency. (2006). Ultraviolet Disinfection Guidance Manual for the Final Long Term 2 Enhanced Surface Water Treatment Rule, EPA 815-R-06-007, Washington, D.C.

van der Graaf, J., de Koning, J., and Tang, J. (2001). "Particle size analysis as a tool for performance measurements in high rate effluent filtration." *Water Science and Technology*, 43(10), 303-310.

Ward, R.L., Bernstein, D.I., Young, E.C., Sherwood, J.R., Knowlton, D.R., and Schiff, G.M. (1986). "Human rotavirus studies in volunteers - Determination of infectious dose and serological response to infection." *Journal of Infectious Diseases*, 154(5), 871-880.

York, D.W. and Walker-Coleman, L. (2000). "Pathogen standards for reclaimed water." *Water Environment and Technology*, 12(1), 58-61.

**WASTEWATER UTILITY**

**Alabama**
Montgomery Water Works & Sanitary Sewer Board

**Alaska**
Anchorage Water & Wastewater Utility

**Arizona**
Avondale, City of
Glendale, City of, Utilities Department
Mesa, City of
Peoria, City of
Phoenix Water Services Dept.
Pima County Wastewater Management
Safford, City of
Tempe, City of

**Arkansas**
Little Rock Wastewater Utility

**California**
Central Contra Costa Sanitary District
Corona, City of
Crestline Sanitation District
Delta Diablo Sanitation District
Dublin San Ramon Services District
East Bay Dischargers Authority
East Bay Municipal Utility District
El Dorado Irrigation District
Fairfield-Suisun Sewer District
Fresno Department of Public Utilities
Inland Empire Utilities Agency
Irvine Ranch Water District
Las Gallinas Valley Sanitary District
Las Virgenes Municipal Water District
Livermore, City of
Los Angeles, City of
Los Angeles County, Sanitation Districts of
Napa Sanitation District
Novato Sanitary District
Orange County Sanitation District
Palo Alto, City of
Riverside, City of
Sacramento Regional County Sanitation District
San Diego Metropolitan Wastewater Department, City of
San Francisco, City & County of
San Jose, City of
Santa Barbara, City of
Santa Cruz, City of
Santa Rosa, City of
South Bayside System Authority
South Coast Water District
South Orange County Wastewater Authority
South Tahoe Public Utility District
Stege Sanitary District
Sunnyvale, City of
Union Sanitary District
West Valley Sanitation District

**Colorado**
Aurora, City of
Boulder, City of
Greeley, City of
Littleton/Englewood Water Pollution Control Plant
Metro Wastewater Reclamation District, Denver

**Connecticut**
Greater New Haven WPCA
Stamford, City of

**District of Columbia**
District of Columbia Water & Sewer Authority

**Florida**
Broward, County of
Fort Lauderdale, City of
Jacksonville Electric Authority (JEA)
Miami-Dade Water & Sewer Authority
Orange County Utilities Department
Pinellas, County of
Reedy Creek Improvement District
Seminole County Environmental Services
St. Petersburg, City of
Tallahassee, City of
Toho Water Authority
West Palm Beach, City of

**Georgia**
Atlanta Department of Watershed Management
Augusta, City of
Clayton County Water Authority
Cobb County Water System
Columbus Water Works
Fulton County
Gwinnett County Department of Public Utilities
Savannah, City of

**Hawaii**
Honolulu, City & County of

**Idaho**
Boise, City of

**Illinois**
Decatur, Sanitary District of
Greater Peoria Sanitary District
Kankakee River Metropolitan Agency
Metropolitan Water Reclamation District of Greater Chicago
Wheaton Sanitary District

**Indiana**
Jeffersonville, City of

**Iowa**
Ames, City of
Cedar Rapids Wastewater Facility
Des Moines, City of
Iowa City

**Kansas**
Johnson County Wastewater
Unified Government of Wyandotte County/ Kansas City, City of

**Kentucky**
Louisville & Jefferson County Metropolitan Sewer District
Sanitation District No. 1

**Louisiana**
Sewerage & Water Board of New Orleans

**Maine**
Bangor, City of
Portland Water District

**Maryland**
Anne Arundel County Bureau of Utility Operations
Howard County Bureau of Utilities
Washington Suburban Sanitary Commission

**Massachusetts**
Boston Water & Sewer Commission
Massachusetts Water Resources Authority (MWRA)
Upper Blackstone Water Pollution Abatement District

**Michigan**
Ann Arbor, City of
Detroit, City of
Holland Board of Public Works
Saginaw, City of
Wayne County Department of Environment
Wyoming, City of

**Minnesota**
Rochester, City of
Western Lake Superior Sanitary District

**Missouri**
Independence, City of
Kansas City Missouri Water Services Department
Little Blue Valley Sewer District
Metropolitan St. Louis Sewer District

**Nebraska**
Lincoln Wastewater & Solid Waste System

**Nevada**
Henderson, City of
Las Vegas, City of
Reno, City of

**New Jersey**
Bergen County Utilities Authority
Ocean County Utilities Authority

**New York**
New York City Department of Environmental Protection

**North Carolina**
Charlotte/Mecklenburg Utilities
Durham, City of
Metropolitan Sewerage District of Buncombe County
Orange Water & Sewer Authority
University of North Carolina, Chapel Hill

**Ohio**
Akron, City of
Butler County Department of Environmental Services
Columbus, City of
Metropolitan Sewer District of Greater Cincinnati
Montgomery, County of
Northeast Ohio Regional Sewer District
Summit, County of

**Oklahoma**
Oklahoma City Water & Wastewater Utility Department
Tulsa, City of

**Oregon**
Albany, City of
Clean Water Services
Eugene, City of
Gresham, City of
Portland, City of Bureau of Environmental Services
Lake Oswego, City of
Oak Lodge Sanitary District
Water Environment Services

**Pennsylvania**
Hemlock Municipal Sewer Cooperative (HMSC)
Philadelphia, City of
University Area Joint Authority

**South Carolina**
Charleston Water System
Mount Pleasant Waterworks & Sewer Commission
Spartanburg Water

**Tennessee**
Cleveland Utilities
Murfreesboro Water & Sewer Department
Nashville Metro Water Services

**Texas**
Austin, City of
Dallas Water Utilities
Denton, City of
El Paso Water Utilities

Fort Worth, City of
Houston, City of
San Antonio Water System
Trinity River Authority

**Utah**

Salt Lake City Corporation

**Virginia**

Alexandria Sanitation Authority
Arlington, County of
Fairfax County
Hampton Roads Sanitation District
Hanover, County of
Henrico, County of
Hopewell Regional Wastewater Treatment Facility
Loudoun Water
Lynchburg Regional Wastewater Treatment Plant
Prince William County Service Authority
Richmond, City of
Rivanna Water & Sewer Authority

**Washington**

Everett, City of
King County Department of Natural Resources
Seattle Public Utilities
Sunnyside, Port of
Yakima, City of

**Wisconsin**

Green Bay Metro Sewerage District
Kenosha Water Utility
Madison Metropolitan Sewerage District
Milwaukee Metropolitan Sewerage District
Racine, City of
Sheboygan Regional Wastewater Treatment
Wausau Water Works

**Water Services Association of Australia**

ACTEW Corporation
Barwon Water
Central Highlands Water
City West Water
Coliban Water Corporation
Cradle Mountain Water
Gippsland Water
Gladstone Area Water Board
Gold Coast Water
Gosford City Council
Hunter Water Corporation
Logan Water
Melbourne Water
Onstream
Power & Water Corporation
SEQ Water
South Australia Water Corporation
South East Water Limited
Sunshine Coast Water
Sydney Catchment Authority
Sydney Water
Wannon Regional Water Corporation
Watercare Services Limited (NZ)
Water Corporation
Water Distribution Brisbane City Council
Western Water
Yarra Valley Water

**Canada**

Edmonton, City of/Edmonton Waste Management Centre of Excellence
Lethbridge, City of
Regina, City of, Saskatchewan
Toronto, City of, Ontario
Winnipeg, City of, Manitoba

## STORMWATER UTILITY

**California**

Fresno Metropolitan Flood Control District
Los Angeles, City of, Department of Public Works
Monterey, City of
San Francisco, City & County of
Santa Rosa, City of
Sunnyvale, City of

**Colorado**

Aurora, City of
Boulder, City of

**Florida**

Orlando, City of

**Iowa**

Cedar Rapids Wastewater Facility
Des Moines, City of

**Kansas**

Lenexa, City of
Overland Park, City of

**Kentucky**

Louisville & Jefferson County Metropolitan Sewer District

**Maine**

Portland Water District

**North Carolina**

Charlotte, City of, Stormwater Services

**Pennsylvania**

Philadelphia, City of

**Tennessee**

Chattanooga Stormwater Management

**Texas**

Harris County Flood Control District, Texas

**Washington**

Bellevue Utilities Department
Seattle Public Utilities

## STATE

Connecticut Department of Environmental Protection
Kansas Department of Health & Environment
New England Interstate Water Pollution Control Commission (NEIWPCC)
Ohio Environmental Protection Agency
Ohio River Valley Sanitation Commission
Urban Drainage & Flood Control District, CO

## CORPORATE

ADS LLC
Advanced Data Mining International
AECOM
Alan Plummer & Associates
Alpine Technology Inc.
Aqua-Aerobic Systems Inc.
Aquateam–Norwegian Water Technology Centre A/S
ARCADIS
Associated Engineering
Bernardin Lochmueller & Associates
Black & Veatch
Blue Water Technologies, Inc.
Brown & Caldwell
Burgess & Niple, Ltd.
Burns & McDonnell
CABE Associates Inc.
The Cadmus Group
Camp Dresser & McKee Inc.
Carollo Engineers Inc.
Carpenter Environmental Associates Inc.
CET Engineering Services
CH2M HILL
CRA Infrastructure & Engineering
CONTECH Stormwater Solutions
D&B/Guarino Engineers, LLC
Damon S. Williams Associates, LLC
Ecovation
EMA Inc.
Environmental Operating Solutions, Inc.
Environ International Corporation
Fay, Spofford, & Thorndike Inc.
Freese & Nichols, Inc.
ftn Associates Inc.
Gannett Fleming Inc.
Garden & Associates, Ltd.
Geosyntec Consultants
GHD, Inc.
Greeley and Hansen LLC
Hazen & Sawyer, P.C.
HDR Engineering Inc.
HNTB Corporation
Hydromantis Inc.
HydroQual Inc.
Infilco Degremont Inc.
Jason Consultants LLC Inc.
Jordan, Jones, & Goulding Inc.
KCI Technologies Inc.
Kelly & Weaver, P.C.
Kennedy/Jenks Consultants
Larry Walker Associates
LimnoTech Inc.
Lombardo Associates, Inc.
The Low Impact Development Center Inc.
Malcolm Pirnie Inc.
Material Matters, Inc.
McKim & Creed
MWH
NTL Alaska, Inc.
O'Brien & Gere Engineers Inc.
Odor & Corrosion Technology Consultants Inc.
Parametrix Inc.
Parsons
Post, Buckley, Schuh & Jernigan
Praxair, Inc.
RMC Water & Environment
Ross & Associates Ltd.
SAIC
Siemens Water Technologies
The Soap & Detergent Association
Smith & Loveless, Inc.
Southeast Environmental Engineering, LLC
Stearns & Wheler, LLC
Stone Environmental Inc.
Stratus Consulting Inc.
Synagro Technologies Inc.
Tetra Tech Inc.
Trojan Technologies Inc.
Trussell Technologies, Inc.
URS Corporation
Wallingford Software
Westin Engineering Inc.
Wright Water Engineers
Zoeller Pump Company

## INDUSTRY

American Electric Power
American Water
Anglian Water Services, Ltd.
Chevron Energy Technology
The Coca-Cola Company
Dow Chemical Company
DuPont Company
Eastman Chemical Company
Eli Lilly & Company
InsinkErator
Johnson & Johnson
Merck & Company Inc.
Procter & Gamble Company
Suez Environnment
United Utilities North West (UUNW)
United Water Services LLC
Veolia Water North America

**Note: List as of 11/10/09**

IWA Publishing's authorised EU representative for General Product Safety Regulations is Diane D'Arras, 15 rue Duret, 75116 Paris, France, e-mail: safety@iwap.co.uk.

Printed and bound by CPI Group (UK) Ltd, Croydon, CR0 4YY
17/09/2026
02218798-0017